WINDOWS & SKYLIGHTS

THE BEST OF
Fine Homebuilding

WINDOWS & SKYLIGHTS

THE BEST OF
Fine Homebuilding

The Taunton Press

Cover Photo: Scott McBride

Taunton
BOOKS & VIDEOS

for fellow enthusiasts

©1995 by The Taunton Press, Inc.
All rights reserved.

First printing: 1996
Second printing: 1998
Printed in the United States of America

A FINE HOMEBUILDING Book

FINE HOMEBUILDING® is a trademark of The Taunton Press, Inc.,
registered in the U.S. Patent and Trademark Office.

Distributed by Publishers Group West

The Taunton Press, Inc.
63 South Main Street
P.O. Box 5506
Newtown, Connecticut 06470-5506
email: tp@taunton.com

Library of Congress Cataloging-in-Publication Data

Windows & skylights : the best of Fine homebuilding.
    p. cm.
    Includes index.
    ISBN 1-56158-127-5
    1. Windows — Design and construction.   2. Skylights —
Design and construction.   I. Taunton Press.   II. Fine
homebuilding.
TT2275.W566   1996
694'.6 – dc20
                                                  95-47082
                                                      CIP

CONTENTS

INTRODUCTION

" **I** AM never able to quell entirely the prickling of apprehension at the moment I lower the blade of my circular saw into a perfectly good roof for the purpose of cutting a gaping hole in it. But once I get going, goggles strapped on, earmuffs dulling the 'chunk' of an old blade ripping through the roofing nails, the excitement always gets me. A hole in the roof dramatically alters the quality of light entering a room." That's builder Tony Simmonds writing about adding a skylight to his home in Vancouver, British Columbia (see p. 102).

Tony's right, you know. Altering the quality of light and admitting a view or a breeze are what windows and skylights are all about. The price we pay for these pleasures is that we complicate the framing of our houses, and we make them less energy efficient and more likely to leak. It's worth it, though.

Collected in this volume are 25 articles from past issues of *Fine Homebuilding* magazine. These articles cover everything from fixing a broken windowpane to choosing between insulated glass with low-E coatings and insulated glass filled with argon gas. In these pages you'll learn about fixing old windows and about making your own windows and skylights.

If you're going to do something crazy like cutting holes in a perfectly good wall or roof, this book can help. It's filled with the expertise of designers and builders who have been there before you.

—Kevin Ireton, editor

Taking a Look at Windows

In a fog about choosing windows? You have to consider glazing systems, window styles and frame materials

by Jefferson Kolle

Part of my job as an editor is to read the hundreds of article proposals that arrive in *Fine Homebuilding*'s mailbox. For me, nothing sends up the red flag of rejection faster than seeing the word *fenestration* in a proposal letter. As in, "We tried to balance the subtleties of the *fenestration* with the implied massing of the built volumes."

Fenestration is a highfalutin word favored by architects and window manufacturers. It comes from the Latin word *fenestra*, which is the opening between the inner ear and middle ear.

Fenestration refers to the choice of windows in a wall or building. When choosing windows for a house, you'll have to decide not only the type of windows—casement, double hung, fixed, awning or slider—but also the kind of glass to use and the material used for the windows' frames. There's a lot of information, but it doesn't have to be confusing.

I've always thought that the most beautiful words—*fenestration* being an ugly one—are those that express ideas simply. So during the course of this article, I'll try to sort through all the window options without using the dreaded F-word.

Glass technology and window styles—Historically, the evolution of window styles has coincided with technological advances in glass-making. American colonists didn't use little panes of glass in their windows because they were trying to make their houses look like Ye Olde Gift Shoppe. Their windows had small panes because the technology of the day prohibited the manufacture of flat pieces of glass bigger than about 4 in. square.

As glass-making technology improved and the price of larger panes of glass got less expensive, windows with larger, fewer panes per window became common. In the early 18th century, it was not unusual for a sash to have 12 panes of glass, or lites, per sash. By the beginning of this century, one-lite windows were getting popular. Float glass, a new process developed in the late

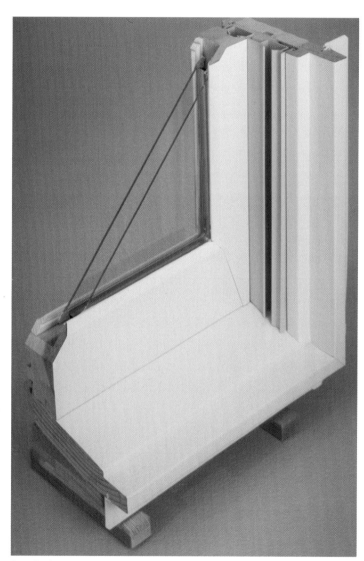

Insulating glass traps air between panes. Air is a poor conductor of heat and thus a good insulator. Insulating glass is made by sealing air between two or three panes of glass. The insulating glass shown here is separated by a stainless-steel spacer that is thermally isolated by polyisobutylene, which reduces conductive losses of heat and cold.

1950s, makes it possible to produce pieces of nearly flawless glass in single-lite window sizes that are limited only by the size of the wall in which you want to install them.

Glass and energy-efficiency—In the 1970s, when everyone's energy-using consciousness was raised, people began to realize that most

windows were nothing more than holes in the wall into which you pour heat. Storm windows had been in use for a long time, but storm windows are often inconvenient and ill-fitting, and they don't adapt readily to all window styles; for instance, you can't open a casement window more than an inch or two if there's a storm window attached to the frame's exterior.

Insulating glass—two panes of hermetically sealed glass separated by a spacer—was invented in the first quarter of this century, but it wasn't until the 1970s that the technology came into widespread use. Air is a relatively poor conductor of heat and cold, and having dead-air space between two pieces of glass works well to improve the efficiency of a window (photo left). The wider the airspace between panes—½ in. to 1 in. is common—the higher the insulating value of the insulating glass. Filling the space between pieces of insulating glass with gases that are less conductive than air—krypton or argon—further improves insulating ability.

Storm windows should not be confused with storm panels. Storm panels are tight-fitting pieces of glass attached semipermanently to a window's sash, where they don't impede the window's operation. Storm panels are not insulating glass; they are not hermetically sealed to the other glass in a sash. And a storm panel mounted over a piece of insulating glass should not be confused with triple-pane glass, which is true insulating glass with three pieces of glass sealed around two spacers.

Low-E glass can work in two ways—At the Massachusetts Institute of Technology (MIT), a group of scientists who later formed Southwall Technologies (1029 Corporation Way, Palo Alto,

Calif. 94303; 415-962-9111) developed the first heat-reflective low-emissivity glass coating. Glass with this coating is known as low-E glass, and it is an option offered by all window manufacturers. The physics of how low-E glass works is beyond the scope of this article, but what it does is relevant and important.

"Imagine a light bulb turned on," says John Meade of Southwall. "When you touch your hand to the light bulb, you are feeling heat through conduction: The heat is passing directly from the bulb to your skin. But if you hold your hand 1 in. away from the light bulb, you still feel the heat, radiant heat this time." Conducted heat can be slowed by insulation: fiberglass batts in a wall or dead-air space between pieces of glass. Radiant heat can be reflected away from the thing being warmed. This reflection is what low-E coatings do.

Low-E glass can work two ways, depending on which way it's facing in the sash. In a cooling environment—one in which you need air conditioning to keep a house temperate—low-E glass can reflect the sun's heat away from a house. In a heating environment, low-E can reflect heat (from your heating system) back into the house instead of having it pass through the windows. The direction in which low-E reflects heat is determined by which glass surface in a piece of insulating glass the coating is applied to.

Southwall developed another technology called Heat Mirror, which is a low-E film that can be suspended in the space between the panes in a piece of insulating glass. Tweaking the chemical makeup of the film's coating allows different types of Heat Mirror films to perform different functions. Some are better at reflecting exterior heat away from a house; others are better at keeping heat inside the house. All Heat Mirror films are superb at keeping ultraviolet (UV) light out of a house. UV light is what fades your furniture and carpets. Insulating glass with Heat Mirror film blocks 99.5% of the UV light that hits the exterior surface of the glass. Low-E insulating glass blocks only around 70% of the UV light.

Armed with insulating glass, triple-pane glass, low-E coatings, Heat Mirror films and exotic gases, window manufacturers and glass companies raced to see who could stuff the most efficiency-increasing components in a piece of insulating glass. A Canadian company, Willmar Windows (485 Watt St., Winnipeg, Man., Canada R2K 2R9; 800-665-8438) has come up with what it calls the R+12 glazing, which is two sheets of Heat Mirror film suspended in krypton gas between two pieces of glass, one of which has a low-E coating. Other companies have different systems for their most efficient glazing systems. As always, high technology and high R-values come with a high price (sidebar p. 12). Willmar R+12 glazing costs more than twice as much as regular insulating glass.

The metal spacer between pieces of insulating glass was a weak link in the efficiency of insulating glass. Metal spacers conduct the heat or cold that the insulating glass was supposed to block. Thus, the center of a piece of insulating glass might have had a pretty good R-value, but all around the perimeter where the pieces of glass

Casement windows swing like doors. Hinged on the side, casement windows swing past the plane of exterior walls. They can act as scoops to catch cooling breezes.

Six-over-six double-hung window pivots for cleaning. A handy feature of some double hungs is that the sash can be tilted into the room so that the windows can be cleaned from the interior.

Three window types ganged in a wall. Window manufacturers can gang different window types together in a wall. Shown here are trapezoidal fixed windows over casements over awning windows.

touched the spacer, the values fell way off. New designs have been developed that improve the cold-edge problem by isolating the metal spacer from the glass or by employing spacers made from nonconductive materials. Every window manufacturer will tout its spacers as the best ones on the market, and currently there is a this-year's-model immediacy in the hype that follows

a window manufacturer's improvement on edge-spacer technology.

Window styles are determined by the direction the sash move—Regardless of the season, my son's favorite bedtime story is *The Night Before Christmas.* About the 20th time I read the tale to him, he stopped me at the line "And threw

Top photo: Pella Corp. Bottom photos: Andersen Windows.

Windows & Skylights 9

Sliding-window sash move horizontally. When they are opened, sliding windows don't protrude past a wall plane, making them perfect for locations where they open onto decks or other outside living areas. Sliding windows usually lift out of their tracks for easy cleaning.

Aluminum windows have thin frames. Because of aluminum's inherent strength, the frames and sash members are typically thinner than windows made of other materials.

open the sash," and he asked me what a sash was. I told him that a sash is one of the two basic parts of a window, the other part being the frame. I went on to explain that the window in the picture book was a casement window, but that I had seen other editions of the story where the window was a double hung. I was about to explain the difference between the casement window in the book and the double hungs in his room when my wife reminded me my son was only 3 years old and told me to keep reading.

All windows are similar in that their two basic components are a sash and a frame. The sash is the part of the window that holds the glass, and the frame is the part that holds the sash. Window types are differentiated by the method and direction in which the sash moves (or doesn't move, as in a fixed window) in the frame.

Double-hung windows consist of two offset sash, mounted one over the other in the frame. The lower sash slides up, and the upper sash slides down. In a single-hung window the upper sash is fixed, and only the lower sash slides.

The sash in a casement window pivots at the side of the frame; casements operate like little out-swinging doors (top photo, p. 9). French casements are two sash hinged on opposite sides of a single frame. Typically, French casements have no center-frame member between the sash.

Awning windows are like casements turned on their side so that the hinges are on the top of the frame, and the sash swing toward the exterior from the top. A hopper window is sort of the opposite of an awning window; the sash in hopper windows swing in, and they are hinged at the bottom. Sliding windows are like double hungs turned on their side; the sash slide horizontally.

Double-hung windows are the most popular—I don't spend a lot of time cleaning windows or even thinking about how easy it would be to clean a window were I so inclined. But window manufacturers give it a lot of thought. At a recent National Association of Homebuilders show, I tried the pivoting action on eight or ten windows on display. The sash on all the new double-hung windows tilted in for cleaning easily. In most cases, all you had to do was release a barrel-bolt-like clip on either side of the sash, and the window tilted right in. Some manufacturers use a compression jamb that allows you to pull the sash toward you easily (bottom left photo, p. 9). Andersen Windows Inc.'s (800-313-4445) double-hung windows even have a mechanism that holds the sash at a convenient angle for cleaning after you tilt it into the room.

One way to differentiate a double-hung window is to refer to it by the number of panes of glass, or lites, in a sash. A 12-over-12 window has 12 lites in the upper sash and 12 in the lower sash. Other common double-hung window patterns are 6-over-6 and 4-over-4. The thin pieces of wood that separate the glass in a sash are called muntins. Anything other than a 1-over-1 sash is referred to as having divided lites.

Divided lites and energy-efficiency—The energy crisis of the 1970s threw a curveball at the divided-lite, double-hung window. There was no denying the importance of energy-efficiency—all anyone had to do was to wait in a gas-station line—and there was no denying the fact that insulating glass was going to save homeowners on their heating and cooling bills.

The first insulated-glass windows typically had one large piece of insulated glass per sash. This design meant no divided lites. Architectural purists were in a quandary: how to get an efficient window and maintain a look they wanted.

Manufacturers came up with the snap-in grille. Snap-in grilles are grids—similar in appearance to a ticktacktoe setup—made of thin pieces of wood or plastic that snap onto the inside of sash and rest against the inside panes of insulated-window glass. There was one problem with snap-ins: They looked as fake as a cheap wig.

Another solution was to make sash that had individual pieces of insulating glass mounted in the rectangles formed by the muntins. But in order to cover the spacers that held apart the two pieces of insulated glass, the muntins had to be uncharacteristically wide. Muntins more than 1½ in. wide were common. Not only was the look inauthentic, but the wide muntins severely reduced the glass area in the sash. Another problem resulted from the increase in metal spacers themselves, which are the most inefficient area of insulated glass. For example, in a six-lite sash, there would be 24 cold edges—four around each piece of glass—instead of a single cold edge around the four sides of the sash's perimeter.

Manufacturers are constantly fiddling with the problem of true divided-lite sash. Some companies do offer sash with individual pieces of insulating glass captured by wood muntins, and some companies offer true divided-lite sash with affixed storm panels. But the more common method of dealing with the problem is to simulate the look. There are several different simulation tactics involving grilles that snap onto the interior of the insulating glass, ones that snap onto the exterior and ones that are sandwiched between the pieces of glass. Most manufacturers offer the previous solutions in combinations: You can order some windows with interior, exterior and sandwiched grilles.

Aside from the obvious energy advantages, another benefit of snap-in grilles (some grilles are permanently attached to the glass) is that some of them are easily removed when it comes time to paint the sash or clean the glass.

Divided-lite sash and simulated divided-lite sash are available from some companies in all their different window styles, not just in their double-hung styles.

Casement windows are hinged on the side—Casement windows were the dominant window style in this country and abroad until the beginning of the 18th century, when the single-hung window came into fashion. Today, casement windows are second in popularity only to double hungs. The hinge side of a casement sash pivots toward the center of the frame when it is opened. When the window is open, you can reach both sides of the glass for washing.

Typically, casements are opened by turning a small crank on the bottom of the frame. Their

Sorting through window ratings

Shopping for an energy-efficient window used to be like trying to buy a used car in a foreign country; you couldn't be sure whether you were being sold a lemon, and you couldn't understand the language. Some window manufacturers spoke of their products' insulation values; some spoke of air infiltration; some spoke of frame conductivity; and some spoke of solar gain. The result was an apples-to-oranges comparison that left you scratching your head.

The NFRC rates windows according to U-values—The National Fenestration Rating Council was started in 1989 to establish a standard energy rating for windows, doors and skylights. NFRC rates windows according to their insulating abilities. Independent testing facilities use computer modeling and actual laboratory tests using windows installed in wall sections to assign a U-value to a window. U-values are the inverse of R-values, so a lower U-value means a higher insulating value.

The NFRC is a nonprofit organization. Its revenues come from selling NFRC stickers with assigned U-values to window manufacturers that put the stickers on their windows.

Critics of the NFRC ratings say there is more to a window's energy-efficiency than its U-value. Air infiltration is important when judging a window's performance. A leaky window with a high insulating value is like a car that gets high gas mileage but leaks fuel all over the driveway.

Canadian ratings also consider air infiltration and solar gain—In terms of window ratings, Canada seems to be ahead of the United States. The Standard of the Canadian Building Code has established an energy-rating (ER) system that incorporates not only insulation values but also air infiltration and solar gain. ER ratings assign a number to a window. Windows are rated on a simple numerical system. A window with a negative ER loses energy, and one with a positive ER contributes energy. A window

with an ER of 0 is neutral in its energy consumption because it contributes as much energy through solar gain as it loses during a heating season. The shortcoming of the Canadian ER system is that it is only applicable to a climate where heating is the predominant energy cost; in the deserts of the United States, where air-conditioning costs are high and heating costs are negligible, solar gain is something to be avoided.

A new rating system is on the way—But take heart. The NFRC is working on a new system that will fill the needs of most all residential consumers. Brian Crooks is an NFRC researcher who works for Cardinal IG (7115 W. Lake St., Minneapolis, Minn. 55426; 612-929-3134), a company that has produced more than 500 million sq. ft. of insulating glass, According to Crooks, the new NFRC system will assign two numbers to a window: a fenestration heating rating (FHR) and a fenestration cooling rating (FCR). The numbers are based on calculations of a window's U-value, its air infiltration and its solar gain. Crooks said that "the numbers represent a percentage of total heating or cooling savings for one window versus another." So if one window is assigned a 10% FHR and another window has a 15% FHR, consumers will be able to tell at a glance that by using the second window they can expect an energy savings of 5% over that of the first window.

As of this writing, the new rating system is awaiting approval by the NFRC. Crooks said the new stickers with both FHR and FCR will be on windows early in 1996.

Between now and the time that the new NFRC ratings appear, your best bet is going to be to use the current NFRC U-value stickers and then spend a fair amount of time reading through manufacturers' catalogs, trying to sort out their convoluted test results. It's probably a good indication that if a manufacturer is forthcoming with its results that its window did pretty well in the tests. If no information is available from the manufacturer, you might ask why.—*J. K.*

ease of operation makes casement windows perfect for locations where sliding sash up or across is inconvenient if not difficult: over a kitchen counter, high up on a wall, etc.

Casement windows lock by means of small levers on the nonhinge side of the frame that clamp the sash to the frame. This clamping action makes casement windows highly resistant to air infiltration, and the harder the wind blows, the tighter the sash is pushed against the frame.

Taller casement windows usually have two lever locks, one above the other, and this upper lever can prove to be an impediment to people in wheelchairs. Peachtree (Box 5700 Norcross, Ga. 30091-5700; 800-477-6544) and Andersen offer optional hardware that allows the lower lever to control both locks. Taller casements made by Pella (102 Main, Pella, Iowa 50219; 515-628-1000) have two locks controlled by a single lever mounted low on the frame. Casement windows can be hinged on either side. And when they are opened, casement windows swing past the plane of an exterior wall. For ventilation they can act as scoops to direct air indoors. Therefore, when you're ordering casement windows, it's important to know the direction of prevailing summer winds in your area because you can order your windows hinged on whichever side takes the best advantage of the natural convection.

Specialty windows—Installing an awning window under a large fixed window can provide ventilation. And placed high on a wall, awning windows can let in air and light while affording privacy. Because they swing outward (top photo, facing page), they can deflect light rain so it doesn't get into a building. Awning windows placed low on a wall can deflect winds hitting the side of a house up into a room.

Windows with horizontally sliding sash are called different things by different manufacturers: gliders, sliders, slide/bys. Both sash slide in a horizontal sliding window, and they usually lift out of their tracks for easy cleaning. On all horizontal sliding windows, the right-hand sash (viewed from the interior) slide on the inside track and the left-hand sash slide on an outside track (top photo, p. 10). The sash slide past each other, but the window can be locked only with the inside sash to the right. Because they don't swing past the plane of a wall like casement windows, sliders are great for locations where you want a window facing a deck or outside space.

Inoperable windows are also referred to as fixed windows, and their shape is, as one catalog says, "only limited by your imagination." Manufacturers have thousands of sizes of fixed windows, and a lot of companies will make a fixed window in any shape you want. Aside from the more common rectilinear fixed windows, most companies have standard sizes of half-round, elliptical and trapezoidal windows. Trapezoidal windows often are installed so that their sloped side is parallel with the slope of a roof (bottom right photo, p. 9).

Because their sash don't open, fixed windows generally resist water and air infiltration well. Window manufacturers spend a lot of time perfecting weatherstripping around operable sash because this location is where water and air tend to invade. A fixed-sash window often is less expensive than an operable window of equal size. For a wall location where it might be hard to reach a window to open it, and where views and light are more important than ventilation, fixed windows can be a money-saving alternative.

Window catalogs are rife with photographs of huge walls of windows, walls that seem to be more glass than drywall. Ganging windows together is a common practice (bottom right photo, p. 9), and window manufacturers sell windows ganged in standard configurations. But windows don't have the strength of a stud wall; there are also wind loads to think about. If you envision a wall of windows for your house, it might behoove you to consult an engineer before you face the likelihood of getting turned down by a building inspector.

Wood windows need maintenance—In the earliest windows, metal was used to hold the glazing in place. By the 18th century wood was the most popular window-frame material, and today, wood windows still command about 50% of the residential-window market.

Until recently, wood has been a plentiful and relatively inexpensive material. And because wood is a poor conductor of heat, wood windows score high on energy-efficiency. But even though all of the parts of a wood window are treated with a preservative prior to assembly, wood windows require maintenance. In order to keep the windows looking good, you're going to have to get out your scrapers, putty and paint every couple of years and have a go at the exterior of a wood window. Some manufacturers will paint the exterior of your wood windows in the factory, and some of their paint jobs come with a good guarantee. Both Weather Shield (1 Weather Shield Plaza, Medford, Wis. 54451; 715-748-2100) and Kolbe & Kolbe Millwork Company Inc. (1323 S. 11th Ave., Wausau, Wis. 54401-5998; 715-842-5666) warrant that their factory-applied coating will last ten years.

Clad-wood windows have a lot to offer—A fellow could get pretty hot and bothered when it comes time to maintain that "warmth and beauty" of his wood windows every couple of years. The perfect answer might be wood windows that are covered on the exterior with either vinyl or aluminum (bottom left photo, facing page). Many manufacturers make clad windows, and they do have a lot of advantages; you get a relatively maintenance-free exterior with a wood interior. Weather Shield even offers windows with oak or cherry on the interior

Although all windows with wood interiors are referred to as clad windows, cutaway photographs in manufacturers' catalogs—and all catalogs show cutaway views—show that there are different ways to clad a window. Some manufacturers attach the cladding by gluing it onto a wood frame. Others have designed their cladding so that it snaps onto a wood frame. Other companies make a vinyl or aluminum frame to which a wood interior is attached. Andersen has a unique system for its vinyl-clad sash: After the wood pieces for the windows are milled, they are covered along their length with a vinyl extrusion. The vinyl-covered pieces then are cut and fit into finished sash. Andersen Windows says that the wood adds strength to the vinyl. Andersen vinyl-clad frames are made by covering an assembled wood frame with an injection-molded vinyl sheathing.

Aluminum-clad windows don't suffer from the energy disadvantages of all-aluminum windows. That's because the heat-conducting properties of the aluminum are broken by the wood interior (photo bottom right, facing page).

The look of real vinyl?—A lot of people think of vinyl windows in the same way they think of vinyl siding: Vinyl is a material appropriate for a trailer, not a house. But vinyl windows are typically a lot less expensive than wood windows. And they never need painting. All-vinyl windows should not be confused with vinyl-clad windows. In the past, some vinyl windows experienced problems. Vinyl expands and contracts at a different rate than glass, and on some windows, repeated thermal cycling caused the vinyl to distort and to pull away from the seals around the glazing. Faulty seals can affect a window's ability to withstand air and water. Dark-colored vinyls absorb more heat than white or light-colored vinyl, and it was the dark-colored windows that had the most problems. One solution to combat the distortion/expansion problem is to form dark-colored vinyl over a light-colored vinyl core.

Vinyl is a brittle material that's 80% salt. In order to make vinyl pliable, a plasticizer is added so that it can be molded. With time, as the plasticizers evaporate, the vinyl gets brittle. New vinyl formulas, called uPVCs (unplasticized polyvinyl chlorides) are supposed to be more stable and resistant to distortion and movement.

An important feature to look for in vinyl windows is welded corners, rather than simple mitered-and-screwed corners. Vinyl corners are welded by heating both sides of a miter until they melt. The melted edges are pressed together; they then cool into one piece.

Vinyl windows are chasing hard at the heels of wood windows in popularity. Vinyl has always been popular for replacement windows, but the market for solid-vinyl windows in new residential construction is growing rapidly. Vinyl windows accounted for almost 16% of the windows sold for new construction in 1994. That number has risen from 3% in 1989.

Aluminum windows have gotten a bad rap—At least that's the opinion of Ralph Blomberg, president of Blomberg Window Systems (1453 Blair Ave., Sacramento, Calif. 95822: 916-428-8060). Blomberg points out that in recent years, some people have steered from aluminum windows because they are not energy-efficient. And some state energy codes have prohibited the use of some solid-aluminum windows. Aluminum is a good heat conductor and, consequently, a poor insulator. But, Blomberg contends, aluminum windows have some advantages, and they remain popular in the temperate West Coast climates. Aluminum windows are typically less expensive than wood windows or vinyl windows, and they shouldn't require much maintenance.

Aluminum can be painted, and a lot of manufacturers offer their products with a factory-applied colored coating. Another advantage of aluminum is that because of its inherent strength, you get a much larger glass area per window size (bottom photo, p. 10) than you can receive with vinyl or wood. And let's face it, you put a bigger window in a wall not so you can see more of the frame and muntins, but so you can look through more of the glass.

Fiberglass is not just for boats and surfboards—Odds are you've never seen a rusty Corvette or a rotten Boston Whaler motorboat.

Awning windows are hinged at the top and swing outward. Awning windows often are grouped below larger fixed windows to provide ventilation in a room. Marvin's Integrity windows, shown here, are wood, clad with fiberglass.

Clad windows should never need painting. The advantage of vinyl-clad windows is that they have a wood interior and a maintenance-free exterior.

Aluminum on the exterior, wood on the interior. Wood's nonconductivity acts as a thermal break for the aluminum, mitigating the energy disadvantages of all-aluminum windows.

That's because both are made of fiberglass, and the stuff is almost impervious to the ravages of weather. Marvin Windows (P. O. Box 100, Warroad, Minn. 56763; 800-346-5128) recently introduced its Integrity line of windows clad with what Marvin calls Ultrex, a fiberglass material. And Blomberg Window Systems has recently introduced a line of all-fiberglass windows.

Fiberglass is not a new material, but only recently was the technology developed to pultrude fiberglass in the thin-walled, complex profiles needed for today's complex windows. Pultrusion is the process in which glass fibers are pulled through a resin bath and then into heated dies that cure the fiberglass.

Fiberglass has advantages as a material for both window frames and sash. Fiberglass expands and contracts at almost the same rate as window glass. This condition is advantageous because different degrees of movement during temperature fluctuations can cause the seals between glass and sash to rupture, which in turn can let in air and water. Other benefits are corrosion-resistance and dimensional stability; a fiberglass window on the south side of a house won't warp and twist. Fiberglass doesn't need much maintenance, although it can be painted. Also, the tech-

nology for pultruding fiberglass currently is limited to straight pieces.

Making an intelligent choice of windows— After you read this article, it's unlikely that you'll know which windows to buy for your next project. And that wasn't the intent any more than a general article about automobiles could tell you what car to buy next.

Before you decide on windows, do what you'd do before buying a car: Go to several dealers; take catalogs home and study them; and go back for another look. Are insect screens optional? Do you want to pay the extra cost of snap-in grilles or spend extra money for more efficient glazing?

Robert Wood at Hurd Millwork Company (575 S. Whelen Ave., Medford, Wis. 54451; 715-748-2011) made a telling comment: "When someone tells me that they don't want to spend a lot of money on windows, I tell them that windows typically account for between 3% and 5% of the cost of a new house. That's not a lot to pay for something you're going to look at and look through every day of your life." □

Jefferson Kolle is a former associate editor for Fine Homebuilding.

Shop-Built Window Frames

Simple joinery, #2 pine and stock sash make inexpensive custom windows

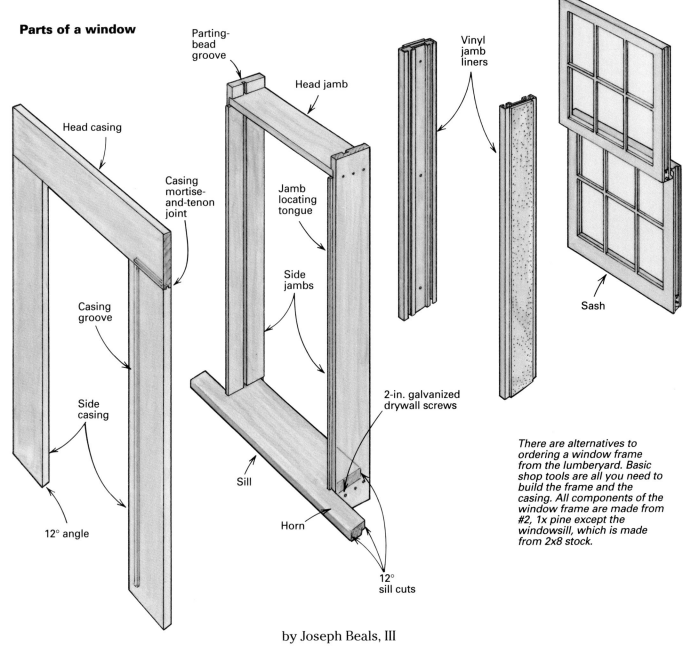

Parts of a window

Head casing

Parting-bead groove

Head jamb

Casing mortise-and-tenon joint

Jamb locating tongue

Casing groove

Side jambs

Side casing

Sill

Horn

12° angle

Vinyl jamb liners

2-in. galvanized drywall screws

Sash

12° sill cuts

There are alternatives to ordering a window frame from the lumberyard. Basic shop tools are all you need to build the frame and the casing. All components of the window frame are made from #2, 1x pine except the windowsill, which is made from 2x8 stock.

by Joseph Beals, III

The window frames in my 1855 house were in varying states of disrepair; some of them needed only a little rebuilding, but a lot of the window frames and most all of the sash were beyond hope. A practical renovation demands modern thermal efficiency at a tolerable cost, as well as some sense of original appearance. These goals can be satisfied by the methods I used for building new window frames.

To attain some degree of thermal efficiency in my windows, I decided to use vinyl jamb liners intended for direct retrofit in existing window frames. If the V word makes the hair on your neck stand up, then you share my first reaction. But in a finished window the jamb liners are quite unobtrusive. A jamb liner is a heavy-gauge, cream-colored vinyl extrusion with full weatherstripping and integral spring counterbalances.

Windows have been built to standard sizes for a remarkably long time; I was a little surprised that I could get replacement sash for my house with dimensions the same as the rotted ones I wanted to replace. But stock window frames are another story.

Most stock window frames (the frames on manufactured windows) are built to fit a contemporary 4½-in. wall and require jamb extensions to suit the thicker walls in an old house. Factory-applied jamb extensions are an option, but the thickness of the walls in an old house often vary considerably, which renders the stock jamb extensions a marginal-to-useless convenience.

Flat exterior casings are a stock option, but they do not match the width of my original casings. Finally, stock windowsills are usually built of thin material, typically 1 in. to 1¼ in. And on some stock frames the sill sticks past the outside edge of the casing only ½ in., which isn't wide enough to add a ¾-in. backband to the casing after the window is installed.

To build my own frames, I use #2 pine for the jambs and casings and kiln-dried 2x8s for the sills (drawing above). I picked the stock to avoid loose knots, pitch pockets, checking and wild grain. Clear pine is an option, but it is very expensive, with no advantage of stability or appearance. Remember, the jambs are not exposed in the finished window, and the sill and the casings are painted on all visible surfaces.

Drawings: Bob Goodfellow

Start with the windowsills

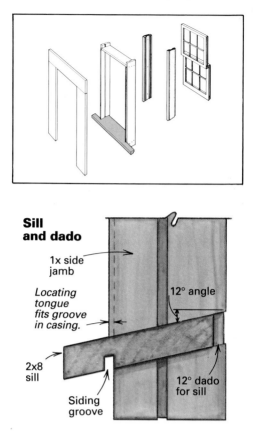

Sill and dado

1x side jamb

Locating tongue fits groove in casing.

2x8 sill

Siding groove

12° angle

12° dado for sill

Windowsills are set at an angle to shed water (drawing above). I used a 12° pitch to match the average angle of the original windowsills. I cut the sill stock a few inches longer than the final size, ripped one edge at 12°, planed it for a smooth finish, then ripped the inside edge to 12°.

When you duplicate old window dimensions, you may find that the inside edge of the sill is short of the inside edge of the jambs. This is not a mistake or an example of Yankee frugality. The inside of the sill need extend only an inch or less in front of the inside of the bottom sash rail, enough to provide nailing for the window stool (the shelflike piece of interior window trim that most people think of as the inside windowsill).

The sill is housed in dadoes angled 12° on the lower ends of the side jambs, with horns that typically are flush with the outside edges of the exterior casings (drawing p. 16).

I think the best method for laying out a sill is to work off a centerline and measure equally to the right and left sides. I mark the location of the inside faces of the side jambs and add the side-jamb dado depth to each side, typically ¼ in. I mark the back of the horns by copying an old sill. In general, I make the horns 2 in. to 2½ in. deep, which gives a reveal of 1 in. to 1½ in. beyond the outside face of the exterior casing. Because the sill is pitched 12°, the actual reveal will be fractionally smaller, but the difference is not worth a trigonometric calculation.

Because both horn cuts are stopped, it doesn't pay for me to make the cuts by machine. I saw out the waste by hand, using a bevel gauge to guide the angle for the horn back cuts (top photo, right). Remember, the back cuts are not critical because they butt against the wall sheathing and are hidden once the siding is in place.

The horn should be flush with the outside edges of the exterior casing. After marking them accordingly, I cut the sills to length on a radial-arm saw. Some early colonial windows have sill horns that extend past the exterior casings, creating an annoying siding condition whereby a clapboard or a shingle would have to be notched around the sill horn with no aesthetic benefit. If you are duplicating windows with this extended-horn detail, consider eliminating it.

Finally, I cut a groove in the underside of the sill where the siding (clapboards or shingles) is tucked as it runs beneath the sill. The back of the groove is flush with the backside of the horns,

Sill cuts. **Because both of the cuts that make the windowsill horns are stopped, the author uses a handsaw to cut away the waste.**

Cutting the siding groove. **A dado blade cuts a ½-in. by ½-in. groove in the windowsill to accept the siding.**

and the groove dimensions are not critical. I make the groove at least ½ in. by ½ in. Making it too tight is a common mistake, second only to forgetting it entirely. A dado blade on the table saw set for width, depth and sill angle makes the groove in one quick pass (bottom photo, above). If you use a router to cut the groove, you can create the pitch angle by taping a shim to the sill, which will lift one side of the router base.

Side and head jambs

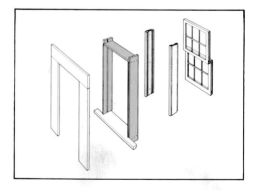

My original windows were built with a rabbet cut on the outside edge of the side jambs. The rabbet leaves a locating tongue that engages a groove on the back of the exterior casings. This is not an essential detail, but it provides positive positioning for the casing and improves the weather-tightness of the joint.

The side-jamb stock is ripped to suit the wall thickness, plus ¼ in. for the rabbet. Add 1⁄16 in. to the actual wall dimension, which will allow the inside of the jambs to stand a fraction proud of the inside finish wall and makes fitting the interior casings much easier. Flush jambs are tolerable if the wall is uniform and dead flat; fractionally recessed jambs are a terrible nuisance.

The length of the side jambs is not critical because the inside frame dimensions are established by the dadoes that house the sill and the head jamb (the top piece in the frame). I always leave an inch or two past the dadoes. Even if it needs to be trimmed later to fit the rough opening, the excess length during construction will protect the short grain between the dadoes and the jamb ends from breaking out. I lay out the dadoes very carefully, working on the left and right side jambs at the same time.

The spacing between the top of the sill and the bottom of the head jamb is defined by the height of the two installed sash (drawing facing page). You can transfer the layout from an old pair of jambs or from a pair of replacement sash.

The width of the dadoes is determined by the stock thickness: For new window frames, that

Angle block. **A scrap of pine cut at 12° acts as a gauge to make the side-jamb sill cut. To cut the opposing jamb, the gauge is switched to the other side of the blade.**

gives 1½ in. at the sill and ¾ in. at the head jamb. A radial-arm saw is ideal for cutting the dadoes because you can see the cut in progress, and a stop block clamped to the fence gives accurate and consistent spacing.

Like most woodworkers, I prefer to leave my radial-arm saw set at 90° rather than setting the left and right angle cuts independently, which increases the possibility for error. To cut the 12° dado for the sill, I make a gauge out of scrap pine. By holding the jamb stock against the gauge, I cut the exact same angle every time (bottom photo, p. 15). To cut an opposing jamb, I simply switch the gauge to the other side of the sawblade. When all the dadoes are cut, I clean out any waste left with a chisel (photo below).

After cutting the dadoes, I plow a parting-bead groove in each side jamb (drawing below). A parting bead is a thin strip of wood that's inserted into the jamb to create separate tracks for the upper and lower sash. A parting-bead groove is unnecessary for a replacement sash that includes a jamb liner, but it ensures the option of historic reversibility. I include the parting-bead groove as a matter of conscience. It is a quick, easy cut, and it will soothe the temper of some future renovator who may already be annoyed by the absence of sash-weight pockets and pulley mortises. I take the groove dimensions and position from an original jamb and plow it with a dado blade on the table saw.

Finally, I cut the rabbet on the outside edge of each side jamb, which creates the locating tongue. The rabbet can be made in one pass with a dado blade on the table saw or with two intersecting cuts to slice off a slender ribbon of waste. The double-cut method is fastest for one window, but a number of windows will justify the time necessary for setting up the dado blade.

In my original windows the head jamb did not have a locating tongue, so I didn't include one. I simply ripped the head jamb to the width of the side jambs, excluding the tongue. The head jamb is the same length as the sill, excluding the horns. It is important that the head-jamb length be accurate; if it isn't, the frame will not be square.

I assemble the frame with galvanized drywall-type screws and construction adhesive. The adhesive is more for waterproofing than for strength, and priming the joints would certainly work as well. I use long screws, at least 2 in., to give adequate grip in the end grain of the sill and the head casing. I also make sure that the sill horns are drawn tight against the outside edge of the side jambs and that the head jamb is flush on both sides. I don't worry about keeping the frame square; even after assembly it will be quite limber and can be squared up as the casings are installed. The final squaring takes place when the frame is installed in the wall.

Making casing

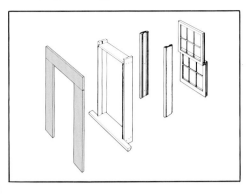

The casings extend ½ in. past the inside face of the jambs. In the original windows this extension served as a stop for the upper sash. For my purposes the stop is a reference edge for the installation of the vinyl jamb liners. It also conceals the edge of the liners. The casing width is equal to the distance from the inside edge of this ½-in. reveal to the outer end of the sill horn.

With the casing stock ripped and planed to its final width, I choose the better face of the board and mark that face with a soft pencil. If there is no clear choice, I orient the heart side toward the weather. Plain-sawn wood tends to cup toward the bark side: With the heart side out, any cupping will be concave in, convex out, which is visually less obtrusive than the other way around.

A groove plowed on the back of the side casings engages the locating tongue on the front edge of the side jambs. You can use a router to cut this groove, but a dado blade on the table saw is fast, accurate and easily adjusted for fit. The tongue should enter the groove easily. Don't try to finesse the joint; this isn't furniture, and a tight spot at the time of assembly will be extremely frustrating.

When all the grooves are cut, I separate the casings into stacks of left and right sides because the remaining operations will define the specific handedness of each piece.

The beveled sill cut on the casing bottoms is best made on a radial-arm saw, but a table saw is fine if you exercise care in controlling the stock to ensure a square crosscut. I use a scrap piece to set up and test the angle, preferably one with the groove on the backside so that it can be set in place on the frame to check the angle. The bottom should sit flat on the sill, but it helps to relieve the back of the angle by a fraction of a degree to keep the front tight. If I can, I make the angle cut with the face side up because both the radial-arm saw and the table saw will tend to chip the bottom of the cut.

A stub tenon on top of each side casing mates with a corresponding mortise on the bottom edge of the head casing (drawing facing page). This joint is perhaps the most tedious part of the frame construction, and although it may appear gratuitous, it actually is quite important. Structurally, it ties the outside top corners of the side casings to the head, which would otherwise float free of each other. But its primary function is to prevent water from working behind the casings.

To lay out the joint, I fit a piece of casing on a frame. I make sure that the casing is well seated

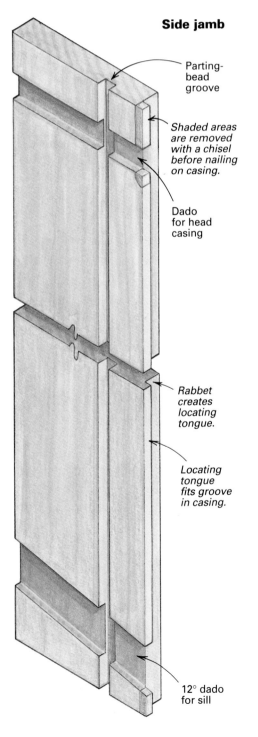

Side jamb

Parting-bead groove

Shaded areas are removed with a chisel before nailing on casing.

Dado for head casing

Rabbet creates locating tongue.

Locating tongue fits groove in casing.

12° dado for sill

Removing the waste. **The author quickly pares away waste with a sharp chisel. Because the orientation of the 12° cuts defines the handedness, left and right jambs are needed for each window.**

on the jamb tongue and that the bottom is tight to the sill. Then I mark a line square across the side casing, ½ in. below the bottom edge of the head jamb. This is the shoulder of the joint, where the bottom edge of the head casing will rest. I mark a second line ¼ in. above the shoulder: This is the end of the stub tenon and the cut-off line for the overall length of the casing pattern (left photo, below).

I cut all casings to length according to the marks. A radial-arm saw makes this easy: I fasten a stop block to an extension fence, butt the bottom of the casing against it and make the cut.

The head casing is the same length as the sill, from horn to horn. After cutting it to length, I mark a line on each end of the bottom edge to show the inside edge of each side casing. I clamp the head casing in a vise, bottom edge up, and use a router with a fence and a ¼-in. carbide spiral bit to cut the two mortises (middle photo, below). The mortises should be a touch deeper than the stub-tenon length to ensure that the head casing sits tight on the side-casing shoulders. To center the mortise, I set the router fence so that the cutter is approximately in the center of the casing edge, then make each cut twice, once from each side and stopping at the line I've drawn on the casing. The first cut removes the bulk of material, and the second shaves off a whisper of waste, leaving a mortise slightly more than ¼ in. wide.

I cut the mortises first because setting up the router cut for a specific width can be very tedious, whereas the stub tenons are easily made to fit a finished mortise.

I cut the stub tenons on the top of each side casing with a radial-arm saw. First I fit a stop block on the fence, using the top of the casing as a reference. Then I set the blade depth on a scrap piece and cut a test tenon. I aim for a snug fit that can be dressed with a sharp chisel if necessary. The stub tenons generally do not justify setting up a dado blade: I just make one or two passes with a standard combination blade, flip the stock over and repeat the cut. Because the typical alternate top bevel (ATB) carbide-tipped blade doesn't make a flat-bottomed cut, a tight fit can be tuned by paring off the peaks of the kerfs.

Because the head-casing mortises are rounded, rather than square at the stopped ends, the inboard ends of each stub tenon need to be relieved. I use a sharp chisel to score the shoulder line across the inboard edge and pop out a little chunk of waste with a quick twist of the wrist.

How it all fits

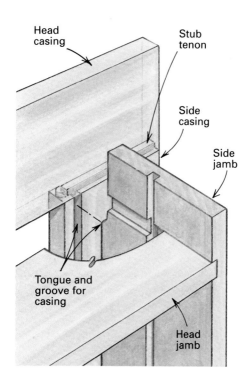

Head casing

Stub tenon

Side casing

Side jamb

Tongue and groove for casing

Head jamb

With chisel in hand, I study the assembled window jambs and sill to see how the jamb rabbets run out to the ends. At the bottom, the rabbets will interfere with the siding installation and should be removed. At the top, the rabbets will interfere with the head casing. I pare off the tongues with a chisel (drawing facing page).

Before installing the casings, I prime the angled bottom cuts and both mortises and tenons at the top. I don't wait for the primer to dry; I place the side casings on the frame, seating the casings fully on the jamb rabbets and making sure the bottom bevel is tight to the sill. For a neat job I drill pilot holes before nailing. I use galvanized 6d common nails to fasten the casings to the jambs, spaced 8 in. to 10 in. o. c. (right photo, below).

When the window frames are finished, they're installed in the usual manner. I tip the frame into the rough opening in the wall and then level the windowsill by shimming under the side jambs on one side or the other. I then use a level to plumb one of the side casings. When the sill is level, and the side casing is plumb, the window will be square. I use 8d galvanized commons to nail through the casings into the sheathing.

Replacement jamb liners are typically sized to suit the stock window and replacement sash, but the jamb liners may not match the sill angle in an old window. To obtain a tight fit at this cosmetically important joint, I ordered jamb liners in the next longer length from Caldwell Manufacturing Company (P. O. Box 92891, Rochester, N. Y. 14692-8991; 716-352-3790) and trimmed them to match the sill with a power miter box. The vinyl saws easily, and the power miter box leaves a clean, accurate cut. The liners attach easily to the jambs with screws provided by the manufacturer. Once the jamb liners are installed, the sash tilt easily into place. □

Joseph Beals, III, is a designer and builder in Marshfield Hills, Mass. Photos by Jefferson Kolle.

Tenon layout. Side casings are laid out for the joint that connects the head and side casings. The left line is the total length of the casing. The other line is the shoulder of the tenon.

Cutting the mortise. The mortise in the head casing is cut with a fence-guided router and a ¼-in. carbide spiral bit. The first pass is made on one side; the second, on the other.

Nailing on the casing. Pilot holes are drilled for 6d galvanized nails. The mortise-and-tenon joint that joins the head and side casings provides weather resistance at the joint.

Double-Hungs Restrung

More than just fishing for sash weights

by David Strawderman

get a lot of calls from clients who want me to replace their windows. Often the old windows are wood sash that have been neglected and are on the verge of falling apart. And sometimes I'm hired to yank out a wall of replacement metal windows that clash with the style of the house. There are many instances, however, when it's best to restore the original windows to working order. This was certainly the case with John and Nina Heaths' recently acquired home. Located in the elegant Hancock Park district of Los Angeles, the house was originally constructed in 1923 by the contractor who built many of the fine surrounding dwellings.

The house was structurally sound, but it had suffered many remodels and lost a good deal of its original refinement. And a lot of things didn't work—like 30 double-hung windows.

Windows with a twist—The windows appeared to be regular old-fashioned double-hung sash of the type popular from the turn of the century to the second World War (drawing below). They were equipped with cords, pulleys and iron weights for counterbalance. A traditional thumb-latch secured the upper and lower sash and a recessed brass finger-pull assisted in raising the lower sash. But unlike other double-hung windows that I had worked on, this finger-pull had a small knob in its center, the significance of which I would soon learn. A closer inspection revealed nice touches. All the interior stops were secured by screws rather than by the usual finish nails. The lefthand parting strips were affixed with flathead screws, a clue that dismantling should proceed from that side.

Most of the windows were painted and caulked shut, but I found one that could still be opened. When I lifted the lower sash, I was surprised to find a split sill containing a recessed wooden bar. A spherical-headed screw protruded from the bar, and when I pulled on it with a pair of pliers the wooden bar revealed itself as the top rail of a concealed window screen. I now understood the function of that knob in the finger-pull: it operated a claw mechanism that clasped the screw protruding from the screen frame. A 90° turn released the screw so the sash could be opened without the screen. The upper sash had a mechanism with a similar device. I was working on a houseful of windows with retractable screens.

It took me two hours to partially dismantle the first window. As it turned out, nearly one man-day was needed to return each window to full working order. Although the retractable screens presented special problems, the steps I took to rejuvenate the windows are similar to those required for all double-hung windows.

Curing sash paralysis—The first step in restoring the windows to full operation was to free each sash. I used a sharp utility knife to cut the paint and caulk between the sash and the stops. This reduces splintering and paint chipping along the intersections of the stops and the sash. Next I used a small flat bar to gently pry the corners of the sash away from the stops. It's a bad idea to pry from the center because old sash rails are often weak, and excessive pressure will crack the glass. For removal, the lower sash must lift enough to clear the sill lip, and the upper sash must move down enough to clear the pulley assembly.

Next I removed the left interior stop. Because these were secured with screws, they were relatively easy to dismantle. Most interior stops, however, are fastened with 6d finish nails, and they should be pried loose with a wide putty knife. I begin near the center of the stop, as the ends tend to be encrusted with thicker layers of paint where they abut the upper stop and sill. Once the middle is free, the ends usually pop out. Then I remove the nails and set the stop aside.

The lower sash of a typical double-hung window can be maneuvered out of its channel once a stop is removed. But these weren't typical windows. They had flashings along their stiles that interlocked with metal channels affixed to the parting strips. As a consequence, I also had to remove the left parting strip before I could pull out the lower sash.

Parting strips are usually unpainted, making their extraction a pretty straightforward process. The parting strip is a single ⅜-in. by ⅝-in. piece of wood let into a groove in the jamb. It is usually secured with three or four

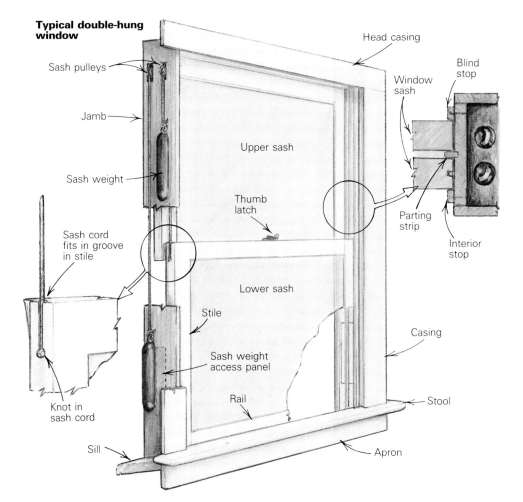

Typical double-hung window

Sash pulleys

Jamb

Sash weight

Sash cord fits in groove in stile

Knot in sash cord

Sill

Head casing

Blind stop

Window sash

Upper sash

Thumb latch

Lower sash

Stile

Sash weight access panel

Rail

Parting strip

Interior stop

Casing

Stool

Apron

finish nails. I located the nails and used pliers to gently rock the strip sideways at each nail (I always place thick cloth or cardboard between tool and wood to prevent marring). Then I slowly pulled it straight out.

The lower sash was finally ready to be removed. Holding it by the rails, I angled its left side into the room and pulled it out of its channel on the right.

Most old double-hung windows have cotton sash cords that deteriorate over the years. Eventually they break and the sash weights drop as far as possible into the wall. Consider it good fortune if you find an intact sash cord, and treat it gingerly. I learned this the hard way—by losing the cord through the pulley. With a helper holding the window, I disengage the cord from the sash and tie the loose end around a short piece of dowel. I repeat for the upper sash. The next step with these windows was the removal of both screen sashes.

Finding the lost weights—Many window frames have access panels, making it a simple process to remove the cover plate and retrieve a weight that has parted company with its sash cord. The weight should be resting on the top of the interior sill extension.

Other window frames, however, have no access panels. Sometimes you can pull the casings off the window frame to get at a lost weight. But that wasn't the case here. The casings were narrow strips that had been plastered in place. As a result, I had to cut access holes in the window jambs (top photo).

Before cutting the holes in the jambs, it helps to calculate the length of the sash weight. Knowing the length tells you how far the access hole needs to extend above the sill to reach the top of the weight. Most weights vary from 4 lb. to 10 lb., and their lengths range from 8 in. to 16 in. To find the length of a sash weight, follow this formula: $\frac{1}{3}$ x sq. ft. of sash = pounds of each weight; $1\frac{1}{2}$ x pounds of weight = length of weight in inches (based on weights $1\frac{3}{4}$ in. in diameter). The poundage is usually stamped in Roman numerals on the side of the weight.

As shown in the photo, I positioned the 2-in. by 14-in. access hole so that most of it is concealed by the inner stop. After drilling pilot holes at the corners of the layout, I used my jigsaw to make the cuts in the jamb.

Once the hole is opened up, the weight may be visible. Use a heavy coat hanger to snag it by the eye. If you don't see it, use the wire as a probe. It is often possible to hear metal on metal and to hook the weight. If these maneuvers are not fruitful, use a small mirror and narrow-beam flashlight tied to a string to search the dark recesses.

Restringing the weights—My pulleys were brass-plated steel, and after 65 years they were still in excellent condition. A little cleaning and a couple of drops of machine oil on each shaft got rid of the squeaks and returned the pulleys to smooth working order. This kind of pulley is held fast by a couple of screws, and

To retrieve the sash weights, the author cut access holes in the window frames. When reassembled, the hole will be plugged and almost completely obscured by the interior stop. Here Strawderman reinserts a sash weight into its chase after attaching a new cord to it.

Changing cords. We pulled this window's casing to reveal the delicate operation. The sash weight is secured with a screwdriver as the old knot is unraveled using needle-nosed pliers.

can be removed from the sash-side of the window. Cheaper pulleys are often press-fit into the jamb and secured from behind with a pin. They are typically serviced by removing the side casings.

I replaced the old cords with the best nylon-reinforced cotton sash cord that I could find. That is Magnolia Sash Cord (Wellington Leisure Products, Inc., 1140 Monticello Rd., Madison, Ga. 30650; 404-342-1916). When replacing sash cord, it's important to avoid a cord with a separate center core—the outer sheath will often wear through prematurely.

The cords to the upper sash are installed first. Before replacing each pulley, I tied a fishing sinker to one end of the sash cord and lowered it through the pulley mortise to the newly cut access hole. Sinker removed, the sash weight is tied to the bottom end of the cord, while the top end needs to be threaded

from the backside through the pulley. A pair of needle-nosed pliers are good for grasping the cord. Before replacing the pulley in its mortise, I tie a half-hitch knot in the end of the cord to keep it from running through the pulley by accident. Now the weight can be reinserted into its channel.

Weights still attached to original cords should be lifted until the eye of the weight is visible through the pulley opening. I pin the weight there with a long, thin screwdriver driven through its eye (photo below). I cut the old cord and replace it with enough cord to reach the sill, allowing an extra foot for attaching it to the sash.

Reinstalling the sashes—Chances are good that the sashes need repainting, and it's best to prep them before putting them back in their frames. This is also an excellent opportunity to clean and sand the frames. I apply a clear wood sealer (The McCloskey Corporation, 7600 State Rd., Philadelphia, Pa. 19136; 800-345-4530) to the unpainted portions of the frames, stops and the edges of the sash.

Make sure you have a helper for the next step. Place the upper sash in its lowered position. Now pull the cord opposite the removed stop until the weight's eye touches the back of the pulley; then let it down about an inch. Determine where the cord fits into the groove in the sash stile, tie a double knot and cut away the excess cord. Have one person hold the cord between the pulley and the sash. Now insert the cord into the slot, place the sash between the blind stop and the parting strip with the sash's edge touching the frame. Carefully release the cord. Pull the other side of the sash into the room and secure the second cord. Raise the sash to the closed position, and repeat the process for the lower sash.

Once the sashes were all back in their frames, I replaced the weight-access panels. I used small wooden wedges to hold them in place, and caulked the saw kerf. After installing the interior stops, I ran a piece of paraffin wax in the channels made by the stops and the parting strips to help the windows run silently and smoothly. □

David Strawderman is a carpenter working in Los Angeles, California.

Sources of supply

If you find yourself in need of replacement parts for old windows, you'll find a number of companies that make or distribute parts. The best guide to who has what is *The Old-House Journal Catalog* ($15.95 from The Old-House Journal Corp., 435 Ninth St., Brooklyn, N. Y. 11215; 718-788-1700). It's a good resource to have on hand for other old-house goodies, too. One of the most comprehensive sources of parts is Blaine Window Hardware, Inc. (1919 Blaine Dr., Hagerstown, Md. 21740; 301-797-6500). Their catalog ($2 to homeowners, free to home-building professionals) is filled with measured line drawings that will help you close in on just what you need.

Building Fixed-Glass Windows

Working on the job site with the tools at hand, you can easily beat the cost of special orders

by Jay Chesavage and Steven Tyler

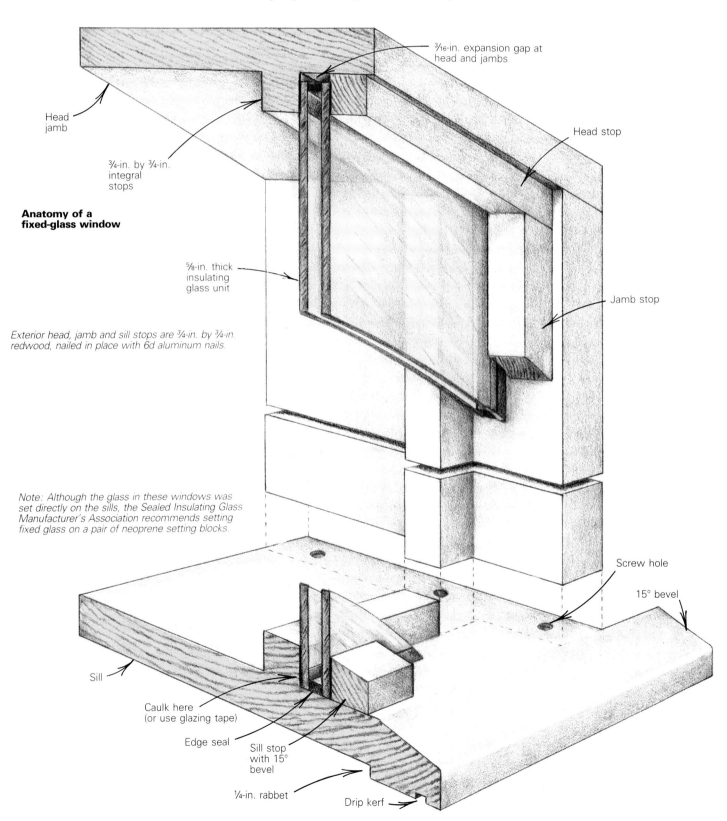

³⁄₁₆-in. expansion gap at head and jambs

Head jamb

¾-in. by ¾-in. integral stops

Anatomy of a fixed-glass window

⅝-in. thick insulating glass unit

Head stop

Jamb stop

Exterior head, jamb and sill stops are ¾-in. by ¾-in. redwood, nailed in place with 6d aluminum nails.

Note: Although the glass in these windows was set directly on the sills, the Sealed Insulating Glass Manufacturer's Association recommends setting fixed glass on a pair of neoprene setting blocks.

Screw hole

15° bevel

Sill

Caulk here (or use glazing tape)

Edge seal

Sill stop with 15° bevel

¼-in. rabbet

Drip kerf

While building a house a few years ago, we experienced the usual number of surpises at material costs and delivery times. One of the worst surprises concerned fixed-glass windows. The home is a passive-solar design, so it includes many such windows (photos below). We reasoned that because these windows were substantially simpler in construction than the wooden casement windows used elsewhere, they would cost less.

A typical wooden casement is a marvel of precision construction, maintaining airtight weatherseals over many linear feet of sash and through years of winter storms and summer heat. In the San Francisco Bay Area, casement windows typically cost $20 per sq. ft. of glazed area, including screens. Although window wholesalers told us not to rely on this approximation, we found it to be generally accurate.

We were shocked to discover that a fixed-glass equivalent would run $35 to $40 per sq. ft. of glazed area—an assembly without moving parts or weatherstripping. Adding insult to injury, the first company we contacted quoted a 14-week delivery time. Other companies promised slightly better delivery times, but couldn't break the $35 per-sq.-ft. price barrier.

We did not consider ourselves window craftsmen. And it remains true that casement windows require so much special tooling that we could not reasonably compete with a production window shop (see the article on pp. 35-39 for more on making casement windows). But Tom Hise, the project architect, convinced us that we could build quality fixed-glass windows in a reasonable amount of time and do so at a price (including our labor, of course) far below the alternatives.

Simple frames—The house is a single-story contemporary with opposing shed roofs and a clerestory. Because the fixed-glass windows follow the roof lines they are trapezoidal. Altogether there are 12 of these windows scattered around the house. We started by laying out all the windows full-scale on the subfloor.

By far, the most important detail in any window is the sill. The sill is the last stopping point for water before it either drips harmlessly off the house, or is pulled destructively into the walls by capillary action. The 15° bevel on the front of the sill directs water away from the building, while the drip kerf underneath the sill guarantees that capillary action won't pull the water into the house (drawing facing page). Milling the jambs and sills from solid stock and incorporating integral stops (rather than using applied stops) similarly prevents water from migrating past the glazing.

The window sills are 1¾ in. thick altogether, with a ¾-in. thick integral stop, a ¾-in. thick center section and a ¼-in. rabbet in the bottom that fits over the rough opening and further discourages water from entering the house. The head and jamb pieces, which are identical to each other in cross section, are essentially sill pieces without the bevel and kerf details, but they're only 1½ in. thick altogether because they don't need the ¼-in. rab-

bet. The jambs and sills are made of #1 clear white pine and were milled on a tablesaw with a dado blade.

All the surfaces that would show after installation were first sanded, and then the pieces were cut to the proper length and angle. The integral stops on the end of each piece had to be cut back (notched) in order to butt the side jambs into the headers and sills (drawing facing page). We cut the sills so that the beveled portion extended past the side jambs on both sides by the width of the exterior trim. The frames were glued and screwed together, then each was laid on top of its respective chalk-line template on the subfloor to check the dimensions.

Setting the frames—After the glue had cured, we painted the window frames, the exterior stop and the exterior trim with two coats of primer. After cutting the exterior trim, we attached it to the jambs and heads with a pneumatic finish nailer, driving 8d aluminum finish nails. We stay away from electroplated galvanized nails because they seem to lose their plating and eventually bleed. If we were hand-nailing, though, hot-dipped galvanized 8ds would have been fine. The nail holes were puttied and sanded smooth.

The finished (but unglazed) frames were then set in the rough openings, plumbed and nailed into place. We use standard flashing details (see *FHB #9*, pp. 46-50), but for good

The window configuration shown above is repeated twice on the rear elevation. So when the window companies quoted a price per sq. ft. for fixed-glass windows that was nearly twice that of the casements, the authors decided to build their own fixed-glass windows.

Building fixed-glass windows is simpler than you might think. After making and installing the window frames, a bead of caulk is run around the interior stops (left) and the double-pane insulating glass is set in the opening (right) and held in place with wooden stops.

measure, we always run a bead of caulk between the top of the trim and the building paper. After installing the siding, we caulked again between the siding and trim.

Fitting the glass—To make sure the insulating glass would fit into our frames, we cut cardboard templates ⅜ in. smaller than the width of the window opening and 3/16 in. smaller than the height and gave them to a glass company. When installing the glass, you can either seal it with glazing tape or with caulking. But if you use caulking, make sure it is compatible with the seal used by the insulating-glass manufacturer. In this case, we used caulk and applied it against the vertical face of each integral stop (photo previous page, bottom left). We followed with a second application on the outside between the glass and the exterior stops. We used

redwood for the exterior stops and nailed them up with 6d aluminum nails. For extra protection, we cut 15° bevels in the sill stops.

We set the glass directly on the sill and have had no problems with it in the two years that the windows have been in place. We've since learned, however, that the Sealed Insulating Glass Manufacturer's Association (SIGMA, 111 E. Wacker Dr., Suite 600, Chicago, Il. 60601) recommends setting fixed glass on a pair of small neoprene blocks (called setting blocks), which help distribute the weight of the glass and prevent water from being trapped behind the glass. SIGMA also recommends drilling a pair of weep holes in the exterior sill stops.

Corner sidelight—At the front entrance of the house we built a large sidelight with two panes of glass meeting at right angles (photo

below). In this case only one header—supported by an exterior wall on one end and by an interior partition on the other—was needed to carry the roof loads. It is conceivable that two headers could be required in circumstances where two load-bearing walls intersect at the window. Several manufacturers of metal connectors make a framing clip for headers that intersect other headers. Because of the unusual glazing detail, it was critical that the rough opening be plumb on either side of the window's corner.

The corner unit was built with the same jamb, sill and head sections as the other windows. First the sill and jamb stock were fabricated as described in the previous sections. Then the sills and heads were mitered and cut to length in matching pairs. Accuracy in cut length was important to guarantee a square opening for the glass. The jambs were cut to matching lengths and the whole unit was assembled near the rough opening.

Before the glue had a chance to set, we placed the unit in the rough opening, aligned the corner of the sill with the corner of the framing, shimmed it level and tacked it in place. We used a plumb bob to align the mitered corner of the head jamb to the identical point on the sill below. Next, the sill was tacked near each jamb and the head adjusted in or out until plumb. Then we stepped back and double-checked that everything was plumb and level.

It was not critical that the corner be exactly 90°. The critical requirement was that the jambs were plumb so that the two panes of glass would meet neatly at the corner. If small adjustments were needed, this was the time to make them. Once everything was plumb and level, we set the nails, puttied the holes and sanded them smooth.

Because of the proximity of the window to the door and to the floor, we had to use safety glass. And to achieve a clean line at the intersection of the glass, we used single panes (¼ in. thick) rather than double-pane insulating glass, which would have made an awkward corner. We installed the glass exactly as before except that we applied a bead of clear silicone between the mating glass surfaces at the corner. After the silicone set, we trimmed off the excess inside and out with a razor blade. The final step was to miter the exterior stops and nail them into place.

Number crunching—When the windows were finished, I calculated how much they had cost us. We paid $4.34 per sq. ft. for the insulating glass, $3.29 per sq. ft. for the tempered glass and billed our time at $40/hr. The cost for all 12 windows and the corner sidelight averaged just under $15 per sq. ft. □

Because it is close to the door and the floor, this corner sidelight had to be made with safety glass. The corner glazing detail is simply two sheets of glass butted together with a bead of clear silicone between them. The drywalled header that carries the roof loads over the window is visible at the top of the photo.

Jay Chesavage is a contractor and engineer in Palo Alto, California. Steven Tyler is a cabinetmaker and carpenter in Concord, California. Photos by the authors.

Making Curvilinear Sash

How to lay out and assemble a semi-elliptical window

by Norman Vandal

In fine Federal and Greek Revival structures, curvilinear windows are common—above pilastered entrances, in gable pediments and incorporated with Palladian windows. Built under roof gables or in other confined areas where conventional rectangular windows couldn't fit, the curvilinear window was a decorative means of providing light to upstairs rooms and attic space. Since I'd recently been commissioned to build such a window for a nearby restoration project, I began to notice many more of the type I'd been asked to create. I became critical, noting some of the small details that separate the fine from the not-so-fine. Fortunately, there are many semi-elliptical windows to be found around Roxbury, Vermont, where I live. I was able to examine different styles and to develop a taste for the most desirable features.

There are probably many ways of producing curved sash, but I could find little information on building this type of window by hand. Avid old-tool enthusiasts, concerned with the function of the tools in their collections, seem to be doing most of the research. As a rule, most craftsmen rely on large millwork companies to produce on the assembly line the items they need, although cabinetmakers sometimes stumble into sash work when making such pieces as corner cupboards or secretaries.

Drawing the semi-ellipse—This is the first step in making curvilinear sash, and will determine the proportions of the window. I learned how to draw an ellipse from Asher Benjamin's

Curvilinear windows were popular features in houses built during the Federal and Greek Revival period. Even today curved sash must be made largely by hand, using traditional techniques of fine joinery.

The American Builder's Companion, an 1806 guide to neo-classical detail and proportions, reprinted by Dover Publications (180 Varick St., New York, N.Y. 10014) in 1969. The method I find most valuable for making a full-scale template is illustrated in figure 1. Using this technique, one can draw concentric ellipses, a necessity for making templates. Note, however, that the resulting curve merely approximates an ellipse, since the points d and d' on the major axis are not foci in the usual sense.

The distance between a and b will be the maximum length of the bottom rail (in my window, 3 ft., with a height of 13½ in.), but you must draw

the entire ellipse to be able to locate the compass points from which the top of the curve is described. The window I made (photo left) is rather elongated in comparison to many old windows I have seen. You can change the proportion of your ellipse simply by experimenting with the compass points d and d'. Moving them closer to the center point in equal increments will elevate the semi-ellipse by the amount of one increment. With a little experimentation, you'll eventually reach a pleasant proportion.

Making the templates—After determining the proportions of your semi-ellipse, begin to construct the templates. For these I use poster board, available in art-supply stores. Try to get board as large as the full-scale sash to avoid taped seams on the template. You'll also need a piece for exterior and interior casing templates.

To construct the templates, you'll need a compass capable of scribing large-diameter circles. For smaller sash, a pair of large dividers will suffice, though large dividers are difficult to locate. You can make dividers of wood, using a nail and a pencil as points and a wing-nutted bolt to secure the joint—a little crude, but functional. My beam compass, an old set of wooden trammel points fastened to a wooden beam, is shown in figure 2. One point is permanently attached; the other, which slides along the beam, can be set in place with a wooden wedge.

Begin by compassing your exterior sash dimensions on the poster board. Then determine the width of the rail (mine generally measure

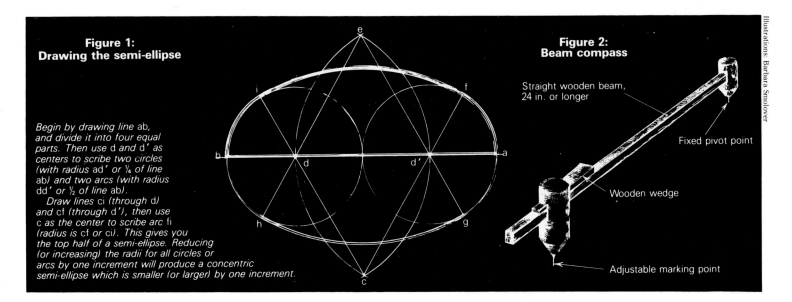

Figure 1:
Drawing the semi-ellipse

Begin by drawing line ab, and divide it into four equal parts. Then use d and d' as centers to scribe two circles (with radius ad' or ¼ of line ab) and two arcs (with radius dd' or ½ of line ab).
Draw lines ci (through d) and cf (through d'), then use c as the center to scribe arc fi (radius is cf or ci). This gives you the top half of a semi-ellipse. Reducing (or increasing) the radii for all circles or arcs by one increment will produce a concentric semi-ellipse which is smaller (or larger) by one increment.

Figure 2:
Beam compass

Straight wooden beam, 24 in. or longer

Fixed pivot point

Wooden wedge

Adjustable marking point

Illustrations: Barbara Smolover

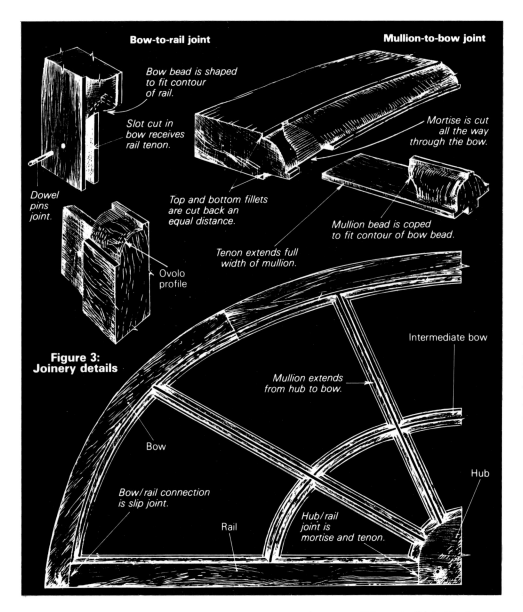

Bow-to-rail joint

Bow bead is shaped to fit contour of rail.

Slot cut in bow receives rail tenon.

Dowel pins joint.

Ovolo profile

Mullion-to-bow joint

Mortise is cut all the way through the bow.

Top and bottom fillets are cut back an equal distance.

Tenon extends full width of mullion.

Mullion bead is coped to fit contour of bow bead.

Figure 3: Joinery details

Intermediate bow

Mullion extends from hub to bow.

Hub

Bow

Bow/rail connection is slip joint.

Rail

Hub/rail joint is mortise and tenon.

Making the mullion-to-bow joint: Once the intersection angle has been marked out on both mullion and bow, the mortises can be drilled out and squared with a chisel. Then the tenon is cut in the mullion and test-fitted in the mortise so the curve can be scribed in the bead, as shown. Use a coping saw to cut the curve and a rattail file or an in-cannel gouge to shape the joint for a snug fit.

1½ in. to 1¾ in.) and the width of the bow. Draw the line representing the top of the bottom rail parallel to the line ab. To draw the inner contour of the bow, you must reduce the radii of the circles with the centers d and d' by the exact width of the bow.

To avoid the disadvantages of short grain, it is best to make the bow out of three pieces of wood instead of one, orienting each to maximize long grain. Draw the lines ci and cf on the template to indicate the position and angle of the joints on the bow. Later, when you're cutting up the template, you can divide the bow into three segments. The sash shown in figure 3 has an intermediate bow, something you may want to avoid in early attempts at making curvilinear sash. If you decide to use an intermediate bow a short distance out from the hub, make it concentric with the curve of the primary bow. An intermediate bow will have one more segment than the number of mullions.

You need not cut templates for the mullions, as they are of straight stock and will be cut to proper length later. But you'll need to decide how many mullions you want in your sash, and to determine what angles they form with bow and hub sections. At this stage you may want to consult old work—the joinery details you are able to uncover in a traditionally made window will prove helpful from this stage onward. In terms of workability and historical accuracy, mullion width should be between ⅝ in. and ¾ in. This will equal twice the width of the bead plus the width of the fillets.

The hub, where the mullions will converge, can be either semicircular or semi-elliptical, but make it large enough to keep the rays from coming into contact with each other, thus keeping the joinery less complicated. The hub must be cut separately, mortised to the rail, and then mortised to receive the mullions.

I usually make my templates ⅛ in. longer than their finished dimensions. You need this extra length when you cut and fit the joints, and any excess can be trimmed off during final assembly. When you have completed the sash templates, make templates for the exterior and interior casings. My exterior casing, which overlaps the sash ½ in. to keep it from falling out, is in two sections, divided by a keystone. Cut out the templates with a stencil knife. I use a surgical scalpel with sharp blades to keep the edges crisp.

Cutting the parts—Your cardboard templates will serve as patterns for laying out all the parts to be cut. I use clear 5/4 eastern pine stock, although mahogany, clear spruce or basswood would serve equally well. Curved cuts for bow sections and the hub can be accomplished with frame saw, bandsaw or saber saw, depending on your equipment and inclination. Remember to cut the mullions a bit long, since some trimming is inevitable with the type of joinery you'll be doing. To join the bow segments, I use a simple butt joint reinforced with dowels—two to each joint—and yellow wood glue.

During the Federal and Greek Revival periods, a complete set of highly specialized tools available from tool-makers and woodworking suppliers provided the joiner with everything needed to make all types of sash. While many of these tools are still around today, most are useful in dealing only with square or rectangular sash. One elusive tool is the sash shave, a cross between a spokeshave and a plane; without it curved sash is practically impossible to make by hand. I could not locate an old sash shave to match any of my planes, so I began to make one from scratch. I quickly realized it was going to take more time and research than I had anticipated (I still haven't finished it), so instead I used a router to cut the molding and the rabbets. You can also use a shaper. The only bits you'll need are a simple ¼-in. bead (with integral stop) and a ¼-in. rabbet (also with integral stop). The bead cutter gives an ovolo molding profile (figure 3), the easiest shape to work with when coped mortise-and-tenon joints have to be cut.

Mortise-and-tenon work—This is where the craftsmanship comes in. The two bow-to-rail joints are the first ones to make. Here you have the option of either mortising a bow tenon into the rail, or cutting a slip joint. I chose the latter because part of the slot and tenon-cutting work could be done on my radial arm saw. The mortise is more secure in terms of joint movement. The procedure for making all sash joints is the same: Lay out the joint first by aligning the parts, mark the stock, and then cut the tenons and mortises. I cut my mortises all the way through, since this makes squaring the cavity easier (you can chisel in from both sides). Check and adjust these for fit, mark where shoulder areas must be coped, and then cope the contoured part of the joint. This is by far the toughest part of the joinery work, and must also be a "check and adjust" operation. I use an in-cannel gouge (a curved gouge with its bevel cut on the inside) and a rat-tail file to shape the curved shoulders of the joint. You may want to use a coping saw to cut the cope into the mullions, but only a gouge can be used on the hub and bow-to-rail joints. Final shaping will consume the extra ⅛ in. or so added to finished length when the parts were first cut.

Once the bow/rail joints are secure, join the hub to the rail (temporarily) and lay out the mullions on the bow/rail/hub assembly to mark where mortises have to be cut. Fashioning the mortises and tenons for mullion-to-rail and mullion-to-hub joints can be made easier if you remember that the tenon extends the full width of the mullion and is made by cutting away the bead or contoured part of the mullion and the bottom fillet. To make the mortise, drill and chisel out the square central portion of the bow or curb between the bead and the rabbet.

Once the mortises and tenons have been cut and fit smoothly, you have to cope the curve in the mullion bead which completes the joint. First seat the mullion tenon in its bow (or curb) mortise, then mark the shoulder cut on the mullion bead, as shown in the photo on the previous page. This has to be done by eye. You can expect some trim work with gouge or file (after you make the cut with a coping saw) to achieve a flush joint, but it's surprising how tight a cope you can cut with a little practice.

When all the mullions have been cut and fitted, a trial assembly is in order. Slip the tenons of the mullions into the bow, and snug up the rail to the bow as you're fitting mullion tenons into the hub. With luck, everything will fit as you planned, but don't be alarmed if a little further trimming is necessary. When you are satisfied with the results, glue up all the joints and reassemble the sash. A peg glued in the two bow-to-rail joints will provide extra strength.

Making an elliptical sash without an intermediate bow, as described above, is the easier and more advisable route if you've never built a curved sash before. If you're incorporating an intermediate bow, as I did, then these bow sections would be cut and mortised into the mullions before final glue-up of the sash.

The jamb—Since all of the old curvilinear window units I observed were in place, I was not able to examine the way in which the jamb units were constructed and framed into the walls. I had to work out my own system, shown in figure 4; you may wish to do likewise. Total wall thickness, sheathing material and stud spacing will determine the width of the jamb and the framing details.

Before you can make the jamb unit, first make the window stool to support the casings. I use 8/4 native white pine, and bevel the stool 10° toward the outside to shed water. (You also have to bevel the bottom of the rail so the rail-to-stool joint is sound.) The stool should be longer than the rail if you want ears at each end. If ears are not part of the design, trim the stool flush once you've attached the jamb.

Although you can make the stool any width, extending it too far increases the amount of weather it will have to endure. Drip kerfs (grooves on the underside of the exterior edge of the stool) allow water to drip free from the building's face. A groove cut to accept the clapboarding under the stool is an aid to both fitting the siding and weatherproofing the unit.

The sides of the jamb unit can be made from two pieces of exterior-grade plywood—I used ¾-in. thick pine stock, although ½-in. plywood will be strong enough if you use more blocking. The elliptical opening in each panel should be a full ³⁄₁₆ in. greater than the final dimensions of the sash on all sides. Once these are cut, attach the blocking that connects both panels. Now you've got a single unit to work on. Invert it to attach the ³⁄₁₆-in. thick jamb sections. They have to be bent to conform to the curve in the plywood frame. I used clear pine and soaked the wood to make it more flexible before fastening it to the ply edges with glue and small brads.

The stool can be attached next, but test-fit the completed sash first—fine adjustments at sash, jamb and stool contact points are more difficult to make once the stool is in place. I used glue and dowel pegs to join the stool to the plywood, but screws will work equally well. Once this is done the exterior casing can be fastened to the exterior side of the frame. It should overlap the bow section of the sash by about ⅛ in. Several small stops, tacked to the jamb on the interior side of the bow, force it securely against the casing.

Allow ½ in. to ¾ in. on both the width and height of the wall opening to shim the jamb unit plumb and level. Fasten the unit by nailing through the exterior casing into the sheathing and frame of the building. Now the siding must be cut to fit the curvature of the casing, but don't despair. You still have the templates, from which you can fit the siding and the interior trim. □

Norm Vandal builds and restores traditional houses. He also makes period furniture.

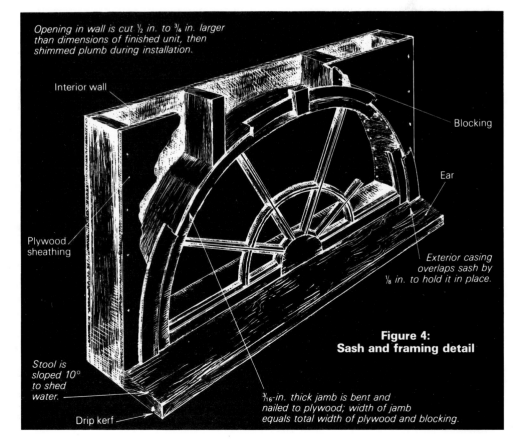

Opening in wall is cut ½ in. to ¾ in. larger than dimensions of finished unit, then shimmed plumb during installation.

Interior wall

Blocking

Ear

Plywood sheathing

Exterior casing overlaps sash by ⅛ in. to hold it in place.

Stool is sloped 10° to shed water.

Drip kerf

**Figure 4:
Sash and framing detail**

³⁄₁₆-in. thick jamb is bent and nailed to plywood; width of jamb equals total width of plywood and blocking.

Making Window Sash

How to do a custom job with ordinary shop tools and a router

by John Leeke

On a historic-restoration project I worked on not long ago, the house's window sash were in poor shape. The original plan for the sash was to repair the worst of them and then replace them all sometime in the future. But before the window work started, the owner decided to have new sash right off. I didn't have enough time to place an order with a custom millworks; so I decided to make the sash in my own small shop, even though it lacks specialized sashmaking machinery. This meant I'd have to match the joinery and molding profile of the originals, and I'd have to work quickly enough to make money on the job without overcharging my customer.

The old sash were hand-made over 150 years ago. One had been without paint for many years, so its joints came apart easily. All I had to do was see how it was made, and reproduce each part. The challenge was to keep track of all those parts, and to make the joints fit properly so the sash would get as much rigidity from its mechanical integrity as from the glue in the joints.

It takes me about 5¾ hours for all the setups needed during a run of sash. The production time for the kind of sash described here on a short run of three or four double units is almost six hours per unit. Considering my time and the cost of materials, the final price was about 20% higher than ordering custom-made sash from the local lumberyard. Not too bad for short-order work that met my specific requirements exactly.

Of course, I could lower these time figures by keeping specialized machinery set up for sash work. If I did, my shop rates would be higher. I'd rather keep my capital expenses low and have more hourly income.

I use white pine for all my sash because it strikes a good balance between machinability and durability. Straight-grain, knot-free wood is essential because the thin, narrow muntins need to be as strong as possible. Also, the outer frame can twist if it's made from wood with unruly grain. I try to use all heartwood, which is stronger and more rot resistant. It's best to cut all the rails and stiles from parts of the board that have vertical grain. These quartersawn lengths of wood are less liable to warp and twist.

John Leeke is an architectural woodworker in Sanford, Maine. Photos by the author, except where noted.

To replace old frames, first remove the exterior casing, as shown above, then remove the interior casing. This exposes the casing nails that hold the jambs to the rough framing. These can either be pulled or cut with a hacksaw blade.

Two kinds of sash—Here I'll describe how I made double-hung sash for jambs that don't have parting strips. So the sash shown in the photos don't have weather stops. *(Parting strips, weather stops and other sash terms are explained on the facing page.)* However, many older sash are made for jambs that have parting strips, and so require meeting rails with weather stops, as shown in the drawing. Meeting rails with weather stops are thicker in section and narrower in elevation than ordinary rails, and are mortised to receive tenons on the stiles, rather than the other way around. If you have to make meeting rails like this, the joinery is the same, except that the stiles are tenoned and the rail is sized to overlap (with

bevel or rabbet) the other rail. You can avoid this trouble altogether, if you wish, by applying the weather stops (with brads and waterproof glue) after the sash are assembled. The instructions that follow are for simple sash.

Sequence of operations—I did all the work on this job with ordinary shop equipment—a table saw, a drill press and a router, which I mounted on the underside of my saw's extension wing. The techniques described in this article can be adapted to produce sash in new construction, casement windows and fixed-glass windows.

I begin by disassembling one of the old sash to determine how it went together, and to get familiar with its decorative and structural details. Then I measure the inner dimensions of the old jamb, and make a drawing of the sash that shows the important features. From the drawing I compile a list that itemizes the parts and tells the dimensions of all the separate pieces.

The sequence of operations in the shop goes as follows: I thickness-plane all the stock (this can be done on the table saw since all the members are fairly narrow), and then cut the tenons and copes on all of the rails and horizontal muntins. Next, I cut the mortises in the stiles and vertical muntins. After this I set up to mold the inner edges of all the frame members on their inside faces, and then I rabbet the same pieces on their outside faces (for glazing). At this point I usually frank the tenons on the rails. Finally, I assemble the sash.

Measurements, drawings, cutting lists—After I take out the old sash, I clean off paint buildup and dirt. If the stiles of the frame are not parallel, I size the sash to the widest measurement and allow a little more time for trimming during installation. If the overall dimensions from sash to sash vary less than ¼ in., I make all the new sash to the largest size and then trim down those that need it after assembly. If the variation is more than ¼ in., I plan to make more than one size of sash. Too much trimming can weaken the frame members.

You can usually make a good guess about the joinery of the original, but if you're doing a precise reconstruction you have to take one of the sash frames apart so you can measure the dimensions of its tenons and mortises.

On my first sash projects I made complete drawings to keep the various parts and joints

Sash anatomy

A basic sash for a double-hung window consists of an outer frame and an inner framework of smaller members that hold the separate panes of glass. The outer frame is made of vertical members called *stiles* and horizontal members called *rails*. The bottom rail on the upper sash and the top rail on the lower sash are called *meeting rails,* and these are often made to interlock when the windows are closed. This interlock can be a mating pair of bevels or an overlap (see section drawings below), and it helps keep out cold drafts. *Plain-rail sash* have meeting rails that lack the interlock feature; their meeting rails simply abut one another. The lower rail of the bottom sash has to be beveled to fit flush against the sill, which should slope toward the outside to shed water.

In the best construction, frame members are held together by wedged through mortise-and-tenon joints; as a general rule rails get tenoned, stiles get mortised. In some traditional sash, though, the meeting rails are rather narrow and so are mortised to house tenons cut onto the stiles. You can use slip joints, but these lose much of their strength if the glue in them fails, whereas wedged through tenons hold firmly even without glue.

The members of the inner framework or grid that holds the glazing are called by several names. I call them *muntins,* though they're variously known as *mullions, sticks, sticking, glazing bars* or just *bars.* Like the outer-frame pieces, the muntins should be tenoned into the rails and stiles, and into one another.

All the frame members—rails, stiles and muntins—are molded on their inner faces and rabbeted to hold glazing on their outer faces. This arrangement requires that tenon shoulders be made to conform to the molded edge of the mortised member.

In traditional sashmaking, there are two ways to shape the tenon shoulder. The first method involves cutting away the molded wood and shaving the shoulder on the mortised piece flat to receive the flat shoulder of the tenon. This means the beads are mitered on both members. The second, and easier, way is to cope the tenon shoulder. Simply stated, a cope is a negative shape cut to conform precisely to the positive shape that it fits up against. —*J. L.*

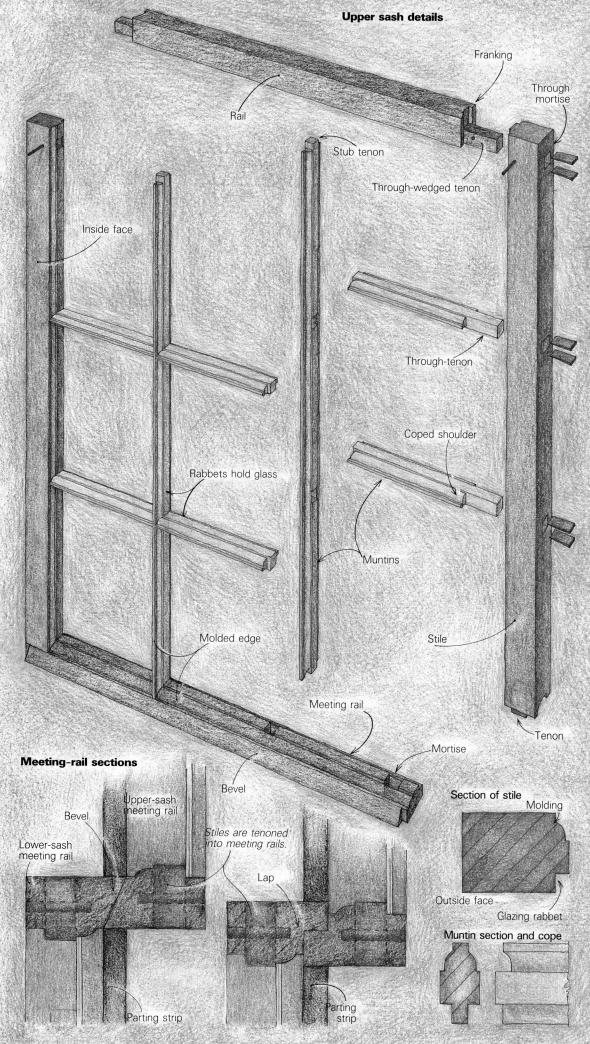

Upper sash details

Franking

Through mortise

Rail

Stub tenon

Through-wedged tenon

Through-tenon

Inside face

Coped shoulder

Rabbets hold glass

Muntins

Molded edge

Stile

Meeting rail

Tenon

Mortise

Bevel

Meeting-rail sections

Upper-sash meeting rail

Bevel

Bevel

Lower-sash meeting rail

Stiles are tenoned into meeting rails.

Lap

Parting strip

Parting strip

Section of stile

Molding

Outside face

Glazing rabbet

Muntin section and cope

Tenoning with a router. With the router mounted on the underside of his saw's extension wing, Leeke removes the correct amount of wood to produce the cheek of a tenon. The stock is fed into the bit with a push gauge; it squares the workpiece to the cutter and keeps the wood from splintering out on the back side of the cut. A shop-vacuum hose pulls chips through a hole in the fence.

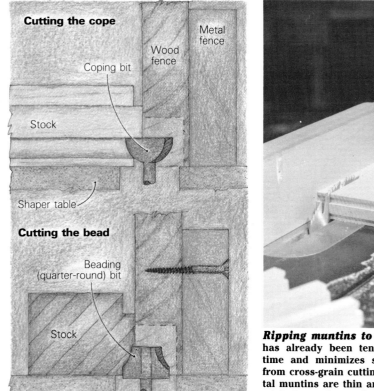

Cutting the cope

Coping bit

Wood fence

Metal fence

Stock

Shaper table

Cutting the bead

Beading (quarter-round) bit

Stock

Ripping muntins to width from stock that has already been tenoned and coped saves time and minimizes splintering and tearout from cross-grain cutting. Because the horizontal muntins are thin and short, a pair of push sticks has to be used.

Rabbeting and molding muntin edges requires the use of two hold-downs. The one attached to the fence holds the stock against the table, and the one clamped to the table holds the work against the fence. Kerfs in the hold-down blocks allow the wood to spring and flex against the stock.

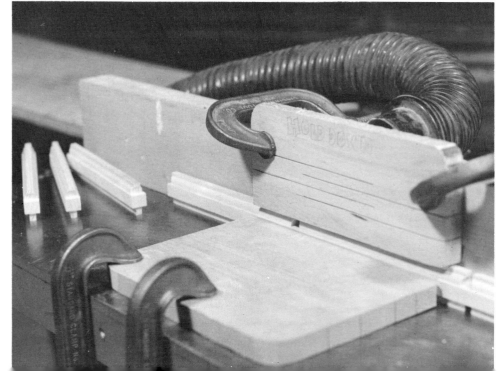

straight in my mind. From these drawings, I made a list of all the pieces I'd need. Each part has its own line, and each line is keyed with a letter to the corresponding part on the drawing. Also on the line is the size, the quantity needed, the name of the part, and its location in the sash.

Router-table joinery—In millwork shops and sash-and-door plants, tenons are cut by single-end or double-end tenoning machines, which cope the shoulder of the joint at the same time they cut the tenon. In smaller custom shops, spindle shapers do the same job. But these are expensive machines, and they take up a lot of space, which I don't have much of. And I was in a hurry. So on this job I improvised a router setup to cut the tenons and copes.

Instead of building an extra table in my already crowded shop, I mounted a router under one of my table-saw extension wings. This arrangement lets me use the saw's rip fence and miter gauge as working parts of the router setup. To keep from drilling needless holes in your saw table, be careful when you lay out and bore the hole in the cast iron for the arbor and cutter and the tapped holes that will let you attach the router base to the underside of the wing with machine screws.

I made a wooden auxiliary fence 2 in. thick and 5 in. high to attach to my saw's metal fence. To suck up chips and dust from around the cut, I bored a hole through the fence and carved a socket to accept the end of my shop vacuum hose (photo top left). This keeps the chips from clogging in the bit during cutting and from building up against the fence.

To guide the workpiece, I use a rectangular push gauge made from a block of pine. By holding the edge of the block against the fence, I stabilize the stock, and square it to the line of cut. Also I notch each corner of the push gauge to each size tenon and to the copes. This way the block backs up the workpiece and keeps the cutter from tearing and splintering the wood as it leaves the cut.

Cutting tenons on rails and muntins—For plain-rail sash there are two tenon lengths—the long through-tenons on the rails and on the stile ends of the horizontal muntins, and the short tenons on the vertical muntins and on horizontal muntins where they are joined to the vertical muntins. It could be that your sash will have a third tenon length for vertical muntins that are joined into the rails, and a fourth tenon length if your stiles are to be tenoned into meeting rails with weather stops. Before setting up to cut tenons, the stock for the rails, stiles and muntins must be surfaced to final thickness and crosscut to finished length. But the stock for the rails and muntins should not at this point be ripped to final width, especially the muntin stock. It's easier and safer to cut the tenons and do the coping on wide boards; it saves time and avoids tearout as well. Remember to mill up some spare pieces for trial fitting, and to be substituted if you ruin good ones. And it's a good idea at this point to set up the hollow-chisel mortiser in your drill press because you'll need to cut some mortises in scrap to test-fit the tenons.

Most tenons are slightly offset from the center of the stock, but because all the framing members are the same thickness, you can set the router bit to cut the tenon cheek on the inside faces of all the pieces, then reset the bit to cut the cheeks on the outside faces. Mark out the dimensions of the two lengths of tenons on a pair of test pieces, and set the bit at the precise height for cutting the face side. This requires careful measuring, for which I use a vernier height gauge.

Calculating tenon length is complicated by the need to cope the tenon shoulder, which in effect lengthens the tenon. This added length equals the depth of the molding profile, and has to be deducted from the length of the tenon. Say your stile width is 1¾ in. and your molding-profile depth is ³⁄₁₆ in.; your through-tenon length before coping will be 1⁹⁄₁₆ in., so you'll set the fence 1⁹⁄₁₆ in. from the farthest point of the bit's cutting arc.

Once the bit is set at the proper height to cut the cheek on the inside-face side of the pieces and the fence is set to cut the longer tenons, you can begin cutting. It's best to make each cut in several passes, even if you're using a large (½-in. or ⅝-in.) carbide-tipped straight bit. You'll get better results without putting an unreasonable demand on the router's motor. Make certain when you make the final pass that the end of the stock is pressed firmly against the fence and at the same time held snugly against your push gauge. Holding the stock this way ensures that the tenon will be the correct length and that the shoulder will be perfectly square.

After the first series of cuts on the rails and on the stile ends of the muntins, you need to set up to make the first cuts on the muntins for the short tenons. To keep from moving the fence and having to set it up again when you cut the cheeks on the outside face of the rails, you can thickness a scrap piece and clamp it to the fence to shim it out from the bit's cutting arc to make a tenon of the correct length.

Once you've made the cuts on the inside face of the muntin stock, you're ready to complete the tenoning by cutting the wood away on the outside face. Leave the shim clamped in place, and reset the router bit to the proper height above the table to make the next cuts. Careful measuring here is critical because your tenons won't fit if the bit is set at the wrong height. So make a cut on one of your spare pieces and trial-fit it in the test mortise. Once you get the bit set correctly, run all your muntin stock through. Then unclamp the spacer from the fence and make the cuts on the outside face of your rail stock and the muntin stock that gets long tenons.

Coping the shoulders—Coping with a router means you have to pattern-grind a matched pair of bits—a concave bit to cut the molding on the inside edges of sash members, and a convex bit to cut the cope on the tenon shoulders (photo above right). The positive and negative shapes of the pair must be perfectly complementary or your joint won't close properly, and will have gaps. The sidebar at right explains how to grind stock high-speed steel bits to get a matched set.

Now you're in a dilemma because you need a molded, rabbeted and mortised stile to test-fit the pieces you've tenoned and are getting ready to cope. The best choice here is to rip a stile to finished width, set up the router to mold the inner edges and rabbet the outer ones according to a full-size drawing of the stile in section. But you're having to perform an operation out of its logical sequence, and that can seem a waste of time. You'll also need a couple of muntin pieces; so rip a couple to width and mold and rabbet them at the same time you do the same to the stile. Whatever you do, don't throw your sample pieces away once you've gone to the trouble to make them. If you ever need them again, you'll save several tedious hours of trial-and-error setup if you have these samples to refer to.

Now that you've got a stile prototype and a couple of muntin samples, chuck the coping bit in the router and set the height so that the top of the cutter just lightly touches the bottom edge of the tenon (stock held inside face down on the table), as shown in the upper drawing on the facing page. Next set the fence to cope the shoulders of the muntins with short tenons. Be conservative when you set up. Make a pass into the cutter, and trial-fit the piece. If the fit is bad, adjust the fence cautiously and try again. Keep making minute adjustments until you get it right. You'll get some tearout on the exit side of the cut because the edge has been molded, but this won't matter for the test piece. Now cope all the shoulders for the short tenons of the muntin stock. Next reset the fence to cope the shoulders of the long tenons.

At this point you should rip the muntins to width (middle photo, facing page). To keep from having to clean up the sawn surface with a plane, I use a sharp planer blade in my bench saw. Because the muntins are thin, I use a pair of push sticks to feed the stock into the blade. Also at this time you should rip the rails and stiles to final width.

Mortising stiles and muntins—I use a ½-in. hollow-chisel mortiser that I keep sharp for this job. You can buy one of these attachments for your drill press at most woodworking-machinery dealers. Be sure to buy a little conical grinding stone that keeps the chisel sharp. A dull chisel will tear the wood on the walls of the mortise and cause nasty splitting out on the back side of the stock. Even a sharp chisel will do some tearing out if you don't back up the cut with a maple block. I use an aluminum plate that's cut out to the precise ½-in. square dimensions of my hollow chisel. The edges of the square hole give positive support to the wood as the chisel exits, and prevent splintering altogether.

To get precise results, each mortise should be laid out with care. It's best if you use a mortising gauge and striking knife, but a sharp

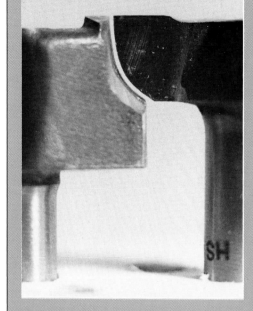

Franking is an operation that removes stock above the tenon to let the shoulder of the rail conform to the profile of the stile. This is easily done with a hollow-chisel mortiser (above). After franking, the waste portion of the tenon is sawn off on the bandsaw (above right). Then the joint can be assembled (right).

Assembly is straightforward. Daub the mortise walls with glue, and tap the frame members together. Then drive in the wedges for the through tenons, and snug up the frame with two bar clamps. The frame shown here is slip-joined because it will be a fixed upper unit and not subjected to the stresses that sliding sash are.

#4 pencil and a try square will do. I clamp an end stop to the drill-press table to ensure that mortises at tops and bottoms of stiles are all the same distance from the ends of the stock.

Molding and rabbeting—The next machining processes mold the inner edges of the frame members on their inside face and rabbet the inner edges on the outside face. First I install the molding bit in the router, adjust its height and set the fence. Having done this already on the test pieces, I use one of them to help with the setup. Then I mold the stiles and rails, after which I clamp hold-downs on the fence and the saw table (bottom photo, p. 28), to run the thin, narrow muntins. Molding done, I cut the rabbets, using the same straight bit that cut the tenons. Again, I clamp hold-downs to the fence and table.

Franking—When you lay out the through mortise-and-tenon joints for the rails and stiles, you'll need to leave at least an inch or more of wood between the end of the mortise and the ends of the stile, or the wood can split out. Therefore, the tenon needs to be an inch or so narrower than the rail is wide. When you cut the tenons with the router, you get a tenon the full width of the rail, so you must cut part of it away to get a tenon the right width and to get the newly exposed shoulder to fit the sectional profile of the stile. Molding and rabbeting the stile produces a proud land (flat ridge) that runs the length of the member, and the new tenon shoulder has to be relieved to accommodate this long, flat ridge.

Relieving the upper shoulder of the tenon in this way is called franking, which I do with the hollow-chisel mortiser (photo top left). What's required is mortising back behind the tenon's shoulder to the width that the land is proud and to a depth that stops at the line of the top of the finished tenon. Then all you do is saw away the waste (photo top right).

Assembly—I really enjoy this part of sashmaking, when all the work finally pays off. Traditionally, joints were put together with thick paint between the parts. I suspect the sashmakers in those days expected the paint to seal moisture out of the joint rather than act as glue to hold it together, but they didn't rely much on the adhesive strength of either paint or glue. They pegged the tenon through the inside face of the stile and wedged it top and bottom from the open end of the mortise.

Most weatherproof glues seem to work well, except for formaldehyde-resorcinol, which can bleed through to the surface after the sash is painted. When all the parts of a sash are fitted together, I snug up the joints with a bar clamp, and wedge the through tenons.

After wedging I check the sash for squareness and then peg the joints. Pegging is especially important for the joints of the meeting rail on the lower sash, because they are subject to a lot of stress during use. If pegs pass through to the outside of the sash, they could let water get into the joint. My pegs stop just short of the outside surface. □

Installing sash

When you're deciding what to do with sash that are in poor shape, you should consider the window as a whole. Examine the frames for deterioration. Windows on the south side of a house are subject to repeated wetting and drying. This can cause large checks or cracks in sills and stiles, so that these members need to be replaced. The north side of the house will be damper because it is in the shade, and you're likely to find rot in the joints of the sill and the jamb stiles.

Jamb and casing—If the sill or jambs need replacing, I usually take the whole window frame out of the wall. I begin by carefully removing the exterior casing and moldings, using a prybar to lever them from the wall. This exposes the space between the jamb stiles and the structural framing (trimmer studs, header and rough sill), which gives me room to cut the nails that hold the frame in place. I use a reciprocating saw or hacksaw blade for this. There also may be nails that hold the interior casing to the frame that will have to be cut.

Once the frame is loose, it can be pulled out at the top and removed from the wall. I try to use heartwood for replacement parts of the jamb and casing because it is more rot-resistant than sapwood. In any case, the parts should be treated with a water-repellent preservative before assembly and reinstallation. Reusing old casing and molding that are still in good shape helps blend the new work in with the rest of the house. If the old jambs are still good, I scrape and sand their inside surfaces so they are flat and smooth for the sash to slide against.

Fitting the sash—To install double-hung sash with the upper sash fixed in place and the lower sash sliding up and open, you first trim the stiles to fit, then glaze and finally install the sash in the jambs.

Start sizing the top sash by planing the edges of the stiles so the sash will fit into the frame without binding. This should be a free-fit, but not so loose it will rattle side-to-side. Put the sash in place and slide it up to the frame header. If the top rail of the sash doesn't fit uniformly flush along the header, it should be scribed to the header with dividers and trimmed to fit. Then fit in the two outer stops, which are strips of wood that lay flat against the jamb stiles and hold the upper sash in place. If the top rail of the bottom sash and the bottom rail of the top sash are made to overlap and form a weather stop, your jambs must be fitted with parting strips. These are strips of wood that are let into grooves, one in each inner face of the jamb, that run the length of the jamb stile and serve to separate the two sash so they don't slide against one another. Fit the parting strips into the grooves in the stiles so they are held in place by compression only. Don't glue or nail them in place.

Next trim the lower sash to fit by planing its side edges until it runs smoothly up and down in the frame. Set the sash in place with the bottom rail resting on the sill. Then scribe the bottom rail to the sill with dividers set to the distance between the top surfaces of the meeting rails. Plane off the bottom rail to the scribed line (photo above right), forming a bevel that matches the slope of the sill. The weather stops should fit tightly together when the bottom rail of the lower sash is against the sill. If too much is planed off the bottom of the lower sash, this fit is lost. So take some care when trimming for this fit.

When the sash are sized to fit, they should be treated with a water-repellent preservative and primed for painting. Do not prime the side edges of the sash. They should be left bare to slide against the stiles.

Glazing, painting and finishing—I usually take sash to the glass shop to be glazed. The glass should be bedded in a thin layer of glazing compound and set in place with glazing points. When complete, the glazing compound should have a neat beveled appearance and not show from the inside.

The sash should have two top coats of paint. I prefer oil-base paints. Run the paint just slightly onto the glass, thereby sealing the glazing from rainwater. Do not paint the edges of the sash that will slide against the frame. When the paint is dry, wash the glass.

To install the window in its jamb, set the top sash in place and slip the parting strips into their slots. Trim any beads of paint that may have dried on the side edges of the lower sash, and test to see if it still slides freely in the frame. When you are satisfied with the way the sash fits, then secure it in place with the beaded or molded inner stop, taking care to use thin brads so as not to split the wood. These stops should be carefully positioned so the sash is free to move but not so loose that it will rattle in the wind. —*J. L.*

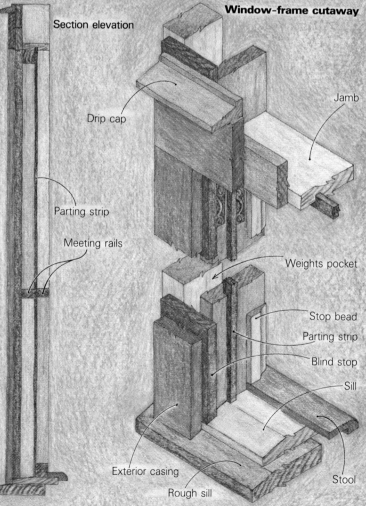

Section elevation

Window-frame cutaway

Drip cap

Parting strip

Meeting rails

Jamb

Weights pocket

Stop bead

Parting strip

Blind stop

Sill

Stool

Exterior casing

Rough sill

Photo this page: Andrew Edgar

Reglazing Windows and Doors

Replacing glass in wooden frames is painless with the right tools and techniques

Sharp as a razor. Shards of broken glass are incredibly sharp, and special care needs to be taken when removing these pieces from the frame.

An old tool with a new job. A bottle/can opener with a sharpened point works great to scrape out the most stubborn old putty that hangs on to the wooden frame.

by William T. Cox Jr.

I recently remodeled an old home 20 miles outside of Memphis. I needed to head back over to the place to finish off a small punch list. Among the items on the list were a couple of panes of glass that needed replacing. The broken panes were in the wooden double-hung windows in the front of the house, plus a pane broken out of a wooden side door. The job wasn't big enough to call in a glass company, so in addition to my regular tools, I loaded my van with some stock sizes of single-strength glass, a box of points and a can of glazing compound.

Gloves may help in removing broken glass—Reglazing is simple as long as the proper tools and techniques are used. The first step is removing the broken pieces of glass as carefully as possible, remembering that the edge of broken glass is sharper than any cutting tool (photo above left). OSHA recommends that gloves be worn at all times when handling glass, but having learned how to handle glass from my uncle, I feel safer being able to feel the glass with bare hands. A lot of the glaziers I've talked to swear by the gummy gloves that are made for handling

glass. The debate over gloves will probably go on forever, but there are no good reasons for not wearing eye protection. Safety glasses or goggles should be worn at all times, especially when removing the glass shards of a broken pane.

After all of the loose pieces of glass are taken out, I remove the old putty. If the windows haven't been reglazed in a while, the putty usually separates from the wood fairly easily. If the old putty is stuck to the wood, a sharp utility knife and small scraper will remove it. I also carry an old bottle/can opener to get rid of hardest stuff (photo above right). The sharpened point of the opener will get into even the smallest cracks.

While removing the glass and old putty, I often come across tiny, flat, diamond-shaped pieces of metal. These things are glazing points, shot into the frame at the factory by a tool similar to a stapling gun. Glazing points hold the glass in the openings of the frames. The putty or glazing compound is meant only to seal the glass from air and moisture infiltration. Without the points the glass would simply fall out over time. Factory points can be saved and put back in, but replacement points are cheap and go in easily.

When the opening is cleaned to the bare wood, I measure it side to side and top to bottom and deduct ⅛ in. in each direction for the glass size. The top panes in the bottom sash of a double-hung window are the exception to this rule. The bottom sash has a slit running along the top stile for the panes to slide into. The top-to-bottom measurement for these panes needs to be ⅛ in. larger than the opening. One-eighth in. of play in each direction should work for all openings that are straight and true. For windows that are badly out of square, I need to figure out whether it's best to replace the broken pane, replace the sash or buy a whole new window.

If more than a couple of panes of glass need replacing, it may be more convenient to take the sash out and repair them on sawhorses. Be aware that removing the sash will create additional work because the sash stops, cords or universal slides will need to be removed, and the inside trim will probably have to be repainted later.

A little light oil makes glass cutting a lot easier—I'm a little leery of hardware stores that sell glass on the side. These places have a con-

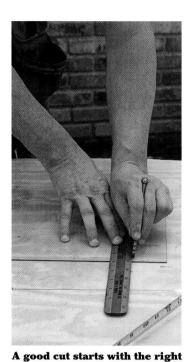

A good cut starts with the right grip. Holding the glass cutter properly is the first key to successful cutting. Apply even pressure and draw the cutter back with a single smooth, continuous motion.

Cutting glass is a snap. After scoring the glass, grip it on both sides of the score mark and apply firm pressure. The pane should break smoothly along the score.

traption that aligns and cuts the glass in one smooth motion. However, I've gotten glass cut on these machines that was so out of square that it wouldn't fit into the opening. A glass shop has all of the equipment needed for precise glass cutting: a flat, cushioned table; a selection of T-squares; oil; and, of course, the best glass-cutting tools. But because there were only a few pieces of glass to install for this project, I decided to cut my own glass on a table made from a piece of plywood on top of my sawhorses. I used an 18-in. ruler for a straightedge.

Glass cutting always goes a lot easier with a very sharp cutting wheel on the glass cutter and a little machine oil. Until the late 1960s, glass cutters were expensive and made with replaceable cutting wheels. Now, with the whole tool costing only a couple of dollars, I keep a new glass cutter handy for each glazing job I do. The machine oil lubricates the cutting wheel, and a little oil laid on the glass just before cutting will prevent a "hot" cut (chunks and shards of glass splintering from the scoring).

Cutting the glass to the right size the first time will save a lot of headaches. Thin strips of glass are difficult to take off with a glass cutter. If I need to fine-tune a glass cut, my belt sander with a 120-grit belt will take off small amounts.

The proper grip on the glass cutter will also help with getting a good cut. Most glass cutters should be held with the handle between the middle and index fingers and the tip of the index finger and thumb gripping the handle just above the cutting wheel (photo top left). If this position feels too odd, hold the cutter like a pencil; either method will give excellent control of the tool.

Start with the cutter wheel at the edge of the glass and press down until the wheel digs in

slightly. Then, with a smooth and continuous motion, score the glass, listening for a long, even, rasping sound as you cut. It's the same principle as hitting the perfect golf shot: It has to be right the first time.

Marks on glass are tough to see and rub off easily, so when I'm ready to cut a piece of glass, I measure with a steel tape and move my straightedge to get the measurement without making marks on the glass. I adjust my measurement to allow for the thickness of the glass cutter, dip the cutter in oil and score the glass once and only once. Scoring the glass more than once may cause it to shatter, or the cut will be ragged at best. Then I pick up the sheet of glass quickly, gripping each side of the score mark with my thumbs on top and index fingers underneath. I apply firm pressure on both sides of the score mark and snap the glass (photo top right). It's necessary to snap glass quickly once it is scored because glass is a supercooled liquid. Scoring the glass disturbs its molecules, and the glass will snap more readily while the molecules of glass are still moving around.

Give the bare frame a coat of linseed oil before glazing—I always dry-fit the pane and then lay it aside. With a rag I rub a good coat of linseed oil on all bare wood surfaces (photo bottom right). Linseed oil is a vegetable oil and one of the main ingredients of glazing compound. Linseed oil refreshes the wood and stops the oil from being wicked out of the newly applied putty. Some folks recommend priming the bare wood with latex paint. I think this does more harm than good because it provides a path between the putty and the wood for air and moisture to get in.

Prep the bare wood with linseed oil. Linseed oil refreshes the wood and keeps the wood from drawing the oil out of the putty.

I dig out a wad of putty about the size of an egg and knead it until it becomes warm and pliable (photo top left, next page). If the original piece of glass was back-bedded, I apply a thin layer of putty on the inside lip of the sash (or muntins) and then press the glass into the putty. Next I install two glazing points along each side of the glass pane (photo top right, next page). I roll the putty between both hands until it resembles a short piece of thick rope (photo bottom left, next page) and work this putty rope into the angle between the glass and mullion. I apply a liberal amount of putty along all four sides with my fingers before I get out my putty knife.

The putty knife should be as clean and as smooth as possible, which is why a lot of them are made with chrome-plated blades. The blade of the knife should also be flexible. With the put-

Warming up the putty. Knead a ball of putty about the size of an egg to warm it and make it soft and workable.

Glazing points are pushed in with a putty knife. With the glass set in the frame, glazing points are inserted to keep the pane in place. Tiny vertical tabs on the points let you use a putty knife to push the points into the wood.

A putty rope is easier to work into the frame. Roll the putty between your hands, and use your fingers to press it into the sash.

A smooth putty knife removes the excess. Drawing a smooth, clean putty knife along each side of the window at the proper angle presses the compound into place, smooths it out and takes off any excess. The proper angle is determined by the height and depth of the rabbet.

ty knife, I press the glazing compound into place as I smooth it out and remove the excess putty (photo bottom right). If the putty knife isn't smooth and clean, it will pull the putty away from the wood. Proper technique is the secret.

I hold the putty knife at a slight angle to the frame, and with a steady backward stroke, I "wipe" the putty into place. The putty knife cuts away any excess, and at the same time, the blade pushes the glazing compound in at the proper angle, which is determined by the height and depth of the rabbet. The best way to smooth any remaining bumps or to close small cracks is to rub the putty lightly with your finger, which builds up a little heat and softens the putty slightly. Rubbing the putty with your finger also brings a little oil to the surface, which I believe aids in the curing of the glazing compound. After I'd fin-

ished glazing all of the new panes in the double-hung windows on the front of the house, I moved on to the side door.

Door lites must be replaced with tempered glass or acrylic—The three-lite, three-panel door had been installed in the 1950s when the house was built, and vandals had recently broken one of the panes. Back in the 50s, door glass did not have to be tempered. However, in 1977 a federal law was passed, mandating that all glass in and up to 1 ft. away from a door be tempered glass or acrylic sheet. It's up to the states, of course, to enforce the law, and here in Tennessee, the law states that anyone buying plain glass must sign a paper stating that the glass will not be used in a door. Before you replace any glass in a door, check your local codes and the

laws concerning door glass, but never use plain glass in a door.

As in most wooden doors, the panes of glass were held in place by wooden stops. With my utility knife I cut the paint away from the seams, or the wooden stops would never have come out in one piece. Tempered glass is impossible to cut, so I had to order the sizes I needed from a local glass company that specializes in tempered glass. After putting a small bead of silicon caulk on the back lip of the rabbet to keep the glass from rattling, I installed the glass and replaced the wooden stops, using 1-in. brads to hold them in place. □

William T. Cox Jr. is a remodel/repair contractor in Memphis, Tennessee, and a frequent contributor to Fine Homebuilding. *Photos by Roe A. Osborn.*

Cranking Out Casements

Using a layout rod and shaper joinery to build outward-swinging windows with divided-light sashes and insulated glass

by Scott McBride

One of the loveliest houses in Irvington, New York, was built in the 1920s by a wealthy philanthropist named Ralph H. Mathiessen. The house commands a breathtaking site on the banks of the Hudson River, overlooking the Tappan Zee. It's an elegant example of the French eclectic style, characterized by decorative brickwork and a steep, slate-covered roof.

Unfortunately, by the time the present owners acquired the house, successive waves of "modernization" had spoiled much of the original detailing. Among other things, casement windows in the third-floor dormers had been traded for aluminum jalousies. I was asked to undo some of the damage by replacing the replacements with true divided-light windows. In addition, the owners wanted to upgrade the thermal efficiency as much as possible, so new glazing had to be insulated glass. None of the new windows we needed were available in stock sizes, and while some of the major window manufacturers now do custom work, the need to save time and money prompted me to build them myself. Over the course of the project, I fabricated sidelights and French doors as well as casement windows using the same basic joinery. But I'll limit the discussion here to casements.

Casement hardware—The job was a first for me, so I familiarized myself with the anatomy of casement windows by reading up on the subject in A. B. Emary's *Handbook of Carpentry and Joinery* (Sterling Publishing Co., out of print) and Antony Talbot's *Handbook of Doormaking, Windowmaking, and Staircasing* (Sterling Publishing Co. Inc., 2 Park Ave., New York, N. Y. 10016, 1980. $8.95). Then I turned to the hardware catalogs to see what kinds of fittings were available.

Casement windows can either swing in or out. Outward-swinging casements (photo right), which I built for this job, are easier to make watertight because rain driven between the sash and the frame drains out rather than in. They are also more convenient to use because they don't project into the room when open, as do inward-swinging casements.

The chief disadvantage of outward-swinging casements is a susceptibility to decay. If left open, even in a light rain, the sash will be soaked. Another problem is that the screen must be placed inside the window, blocking access to the sash. At one time screens were

hinged so that you could get at the old sliding-rod hardware used to open and close the windows. Modern casements open and close by means of a worm-gear crank mounted on the stool. Only occasional access to the sash is required, so the screens can be snapped in place.

Sliding-rod and crank-type hardware (known as "casement operators") is still available from H. B. Ives (a Harrow Co., P. O. Box 1887, New Haven, Conn. 06508). The crank-type operator is by far the more practical, so I incorporated it into my design for the Mathiessen house windows. Although we wanted brass to match the rest of the hardware, Ives makes the operators only in painted steel.

Modern casements have a lever-type lock that operates with the screen in place, but my

customers wanted the look of traditional brass casement locks (called "fasteners"), which were available from Ives. Since the worm gear of a crank-type casement prevents the window from being forced open, even when unlocked, the main purpose of the fastener is to provide a tight weather seal.

Casement frames—Fortunately, many of the original cypress window frames in the dormers were reusable. Most of these had simply been boxed around to make a finished opening for the aluminum jalousies. Where the frames were damaged beyond repair, I removed them and made new ones.

The new side and head jambs have two rabbets each—one on the outside for the sash

Built in the 1920s, this house suffered subsequent remodeling that saw the original casement windows in the dormers replaced by aluminum jalousies. But the wheel of architectural fashion has now come full circle and the casements are back; this time with double-pane insulated glass.

and one on the inside for the screen. These rabbets could have been cut into thick stock, but I found it easier and more economical to attach a ½-in. thick stop to ¾-in. thick jambs.

For the window sills, I decided to use cypress. Although some folks say the cypress available nowadays has grown too fast to contain the resin needed for decay resistance, I gave cypress the benefit of the doubt. The good condition of the original casement frames argued strongly in its favor. I cut a pair of shallow rabbets in the sills, known as "double sinking," to help to keep rain from blowing under the sash (drawing below). A drip groove on the bottom of the sill prevents water from being drawn under the sill by capillary action.

Designing for insulated glass—By far the trickiest part of the job was designing and building the new sash. Switching from ⅛-in. single-pane to ½-in. double-pane insulated glass required an overhaul of the original sash design. Unlike single-pane glass, which is sealed with a thin bead of putty, insulated-glass units are typically held in place by moldings because the solvents in putty will attack the rubber seal around insulated glass. Consequently, a rabbet for insulated glass must be deep enough to accommodate both the increased thickness of the glazing and the thickness of a wood molding. Although the original casement sashes in the Mathiessen house were a hefty 2¼ in. thick, the rabbet for glass was only ¾ in. deep—not enough to hold insulated glass. For this reason I made replacement sashes for some of the remaining original wood casement windows, as well as for those that had been switched to aluminum.

At my local supplier, I had a choice of either 8/4 or 12/4 rough stock. I decided to build the new sash from 8/4 stock planed to 1¾ in. On the outside of the sash, I cut a ⅝-in. deep cove-and-bead profile on a shaper. This profile is called *sticking* because it is worked directly on the sash—"stuck" rather than applied. The glass rabbet took up the remaining 1⅛-in. thickness of sash: ½ in. for the glass and ⅝ in. left for an applied molding similar in profile to the cove-and-bead sticking.

Muntins make up the inner framework, or grid, that holds the glazing in a divided-light sash. On this project, the width and depth of the muntins had to be carefully considered. I was doubling the weight of the glass, so stronger muntins would be needed. In addition, the applied molding had to be wide enough to hide the ½-in. deep rubber seal that separates the two panes in an insulated-glass unit. But if the muntins were too wide, the sash would look more like a dungeon grate than the delicate tracery I wanted. I set-

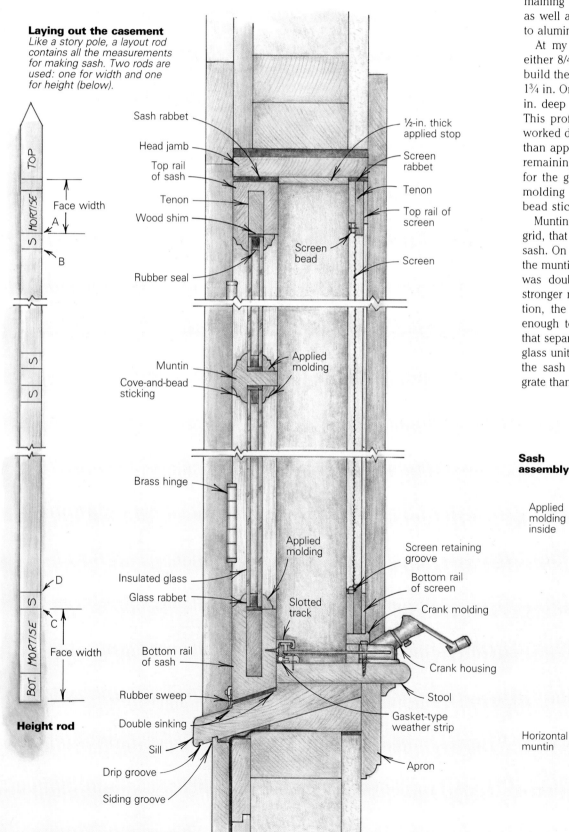

Laying out the casement
Like a story pole, a layout rod contains all the measurements for making sash. Two rods are used: one for width and one for height (below).

Sash rabbet
Head jamb
Top rail of sash
Tenon
Wood shim
Rubber seal
Muntin
Cove-and-bead sticking
Brass hinge
Insulated glass
Glass rabbet
Bottom rail of sash
Rubber sweep
Double sinking
Sill
Drip groove
Siding groove

½-in. thick applied stop
Screen rabbet
Tenon
Top rail of screen
Screen bead
Screen
Applied molding
Applied molding
Slotted track
Screen retaining groove
Bottom rail of screen
Crank molding
Crank housing
Stool
Gasket-type weather strip
Apron

Height rod

TOP
S MORTISE
Face width
A
B
S
S
S
D
S
MORTISE
C
Face width
BOT.

Sash assembly

Applied molding inside
Stile
Horizontal muntin
Rail
Vertical muntin

tled on a muntin width of 1½ in. This would accommodate a ½-in. wide flat face down the middle, with ¼ in. sticking on either side.

Choosing materials—I chose white pine for the sashes, jambs and trim. This old standby is stable and easily worked, and is fairly decay resistant. Sash work requires a lot of short pieces, so I figured I could cut around the knots in Grade 2 material without too much waste. I saved the long, vertical-grained pieces from the edge of each board for stiles and used the less stable flat-grain stock from the middle of the board for the rails and muntins. All this selective cutting took time though, and I've since found it cheaper to buy clear sugar pine in the first place.

I got my insulated glass from a local glazier, who gets it from a wholesale fabricator. Quality control was less than rigorous, however, and I returned a lot of defective units. The defects included corners out of square and ruptures in the rubber seal. I was particularly careful to root out the latter—I've no desire to be replacing fogged-up units a year or two down the road.

Rod layout—Like a story pole for a house, a rod is a layout stick containing all of the pertinent measurements for making a sash. You need two rods—one for height and one for width—and laying them out is the single most important step in sash-making. It requires the builder to think through the entire cutting and assembly process before any sawdust flies.

I began the layout of the height rod by cutting a piece of 1x2 a few inches longer than the height of the sash. I cut one end square and marked it with an arrow and the word "bottom." Starting from this end, I laid off the overall height of the sash, marking this line with an arrow and the word "top."

From the top, I measured back a distance equal to the face width of the top rail, made a mark (point A, drawing facing page), then added the sticking width and made a second mark (point B). I repeated this for the bottom rail (marking points C and D).

I measured the distance from C to A and deducted from this measurement the combined face width of all horizontal muntins. For the sash in this example, the combined thickness is 1½ in. (3 x ½ in.). I divided the balance by the number of lights the sash has along its length (four). The result was the distance from rail to muntin and muntin to muntin, measured on the inside of the sash. This could also be expressed as the distance from shoulder of glass rabbet to shoulder of glass rabbet. After determining the measurments, I marked them on the rod. Then I added the sticking width (½ in.) to both sides of each muntin, marking it with an "S."

Because hinged casement windows are subjected to a racking force resulting from their own weight, they are best made with mortise-and-tenon joints, rather than with the bridle joint sometimes used for double-

hung sash. So the next step in preparing the height rod was to lay out the height of the mortises to be cut in the stiles. Although I wanted maximum surface area in the joint for gluing strength, I was careful to leave enough wood (½ in.) between the mortises and the ends of the stile to prevent the wood from fracturing.

I labeled the mortise heights on both ends of the rod. The smaller mortises—those for the stub tenons in the horizontal muntins—had essentially been laid out already because their height is the same as that of the muntin face width, which I had marked earlier.

When I'm building larger casements or divided-light doors, I usually have a through-tenoned horizontal muntin in the center, tying the stiles together and making the sash more rigid. I didn't think that was necessary here. The casement sashes were small enough that the stiles and rails provided the necessary structure. The muntin configuration consists of seven pieces: one through-tenoned vertical muntin (or sash bar), and six stub-tenoned horizontal muntins (sash assembly drawing, facing page).

In traditional sash-making, the stub tenon extends only as deep as the sticking

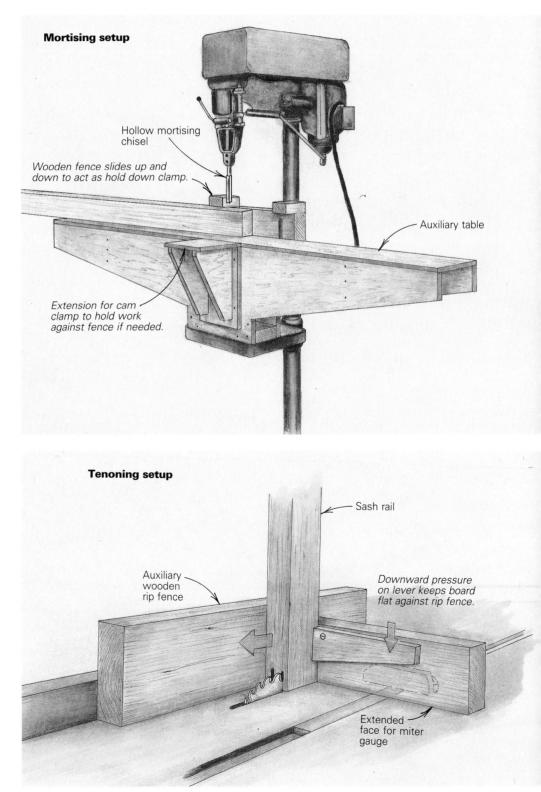

Mortising setup

Hollow mortising chisel

Wooden fence slides up and down to act as hold down clamp.

Auxiliary table

Extension for cam clamp to hold work against fence if needed.

Tenoning setup

Sash rail

Auxiliary wooden rip fence

Downward pressure on lever keeps board flat against rip fence.

Extended face for miter gauge

(for more on traditional sash-making, see pp. 26-31). Because I needed a bigger rabbet to make room for the insulated glass, I had less stock to house the stub tenons. Consequently, they had to reach somewhat further, extending into the face width of the adjoining member. At the vertical muntin, stub tenons would be reaching in from both sides. Because the face width of the vertical muntin is only ½ in., the stub tenon length had to be a little less than half that, or ³/₁₆ in.

To finish the height rod, I labeled it with the word "height" on the top, then cut a point on this end to avoid confusing it with the squared-bottom end, which would be used to register the rod on the stock.

Layout of the width rod was essentially the same as for the height rod. I used through-tenons on the rails, so their length was determined by the overall sash width.

Fabrication—The first step in the actual making of the sashes was to size each piece to rough width and length. Then I jointed the pieces and ripped them to finish width. I saved the thin off-cut strips from the final ripping operation for shims. Varying in thickness from paper-thin to ⅛ in., they came in handy later for positioning the insulated-glass units in the sash.

After crosscutting each piece to finish length, I laid out the mortises. To speed this up, I transferred the mortise heights from the rod onto one of the sash pieces, lined that piece up with all the others like it and squared across the lot of them. Although you can cut mortises with a hand-held electric drill and clean them up with a chisel, I find it a lot faster and easier to use a ½-in. hollow mortising chisel chucked in a drill press. I

didn't need to mark the width of the mortise because it was controlled by the fence setup on the mortiser. My vintage Delta home-shop drill press is a precise but willowy machine. To enable it to handle the long, heavy stiles, I built an auxiliary table out of plywood that extends about 4 ft. on both sides of the drill press (top drawing, previous page). A short wooden fence at the back of the table slides up and down on a pair of bolts, doubling as a hold-down clamp when the wing nuts are tightened. I added a small extension at the front of the table to accommodate a cam clamp for holding the work against the fence, although in most cases I found thumb pressure to be adequate for this.

I cut the tenons on a tablesaw. The combination of a high auxiliary rip fence and a miter gauge enables me to do fast, accurate tenoning without a $200 jig. Before cutting the tenon cheeks, I screwed a wood extension to the face of the miter gauge and attached a wooden lever to that (bottom drawing, previous page). The lever is made so that slight downward pressure on one end will hold the board I'm tenoning snug against the fence.

Coping and sticking—I'm lucky to have two shapers in my shop—a light-duty shaper with a ½-in. spindle and a two-speed heavy-duty shaper (both were made by Rockwell, now Delta International Machinery Corp., 246 Alpha Drive, Pittsburgh, Pa. 15238). In an old manual called *Getting the Most Out of Your Shaper* (originally published by Rockwell, reprinted by Linden Publishing Co., 3845 N. Blackstone, Fresno, Calif. 93726), I read up on the use of shapers for sash-making. Then, armed with matching cove-and-bead molding and cope cutters, I had at it.

In order for the tenoned parts of the sash to fit snugly against the mortised parts, they have to be notched to fit over the sticking (drawing below). This notching process is called coping, and it's the trickiest operation in sash-building, especially when the joinery is mortise and tenon. This is because a through-tenon on a rail or muntin won't clear the top of a standard shaper spindle sticking above the cope cutter. The solution is to use a stub spindle, which is shorter than a standard spindle and uses a countersunk machine screw instead of a nut to hold down the cutter. This allows the tenon to ride just over the cutter (drawing below).

Unfortunately, you need counterbored cutters to use with the stub spindle, and these don't seem to be available in carbide (probably because most sashes are dowelled these days, so cutting around a tenon isn't necessary). As it turned out, the Delta high-speed steel #09-137 counterbored cope cutter costs more than a similar cutter in carbide, but the carbide version isn't counterbored. A machinist counterbored my carbide cove-and-bead cope cutters (#43-915 and #43-916) so that I could mount them on the stub spindle to cope the shoulders of pieces with through tenons.

To form the stub tenons on the muntins, I used the same cutter on a regular spindle in conjunction with a spacer collar and a ½-in. straight cutter with shorter radius. This way I could cope the ends and form the tenons in a single pass.

Some sash builders cope the ends of all the muntins at once by cutting the coped shape across the end of a wide board and then ripping the individual muntins from it. I ripped each one to size first, and then coped them individually, using a sliding jig that Del-

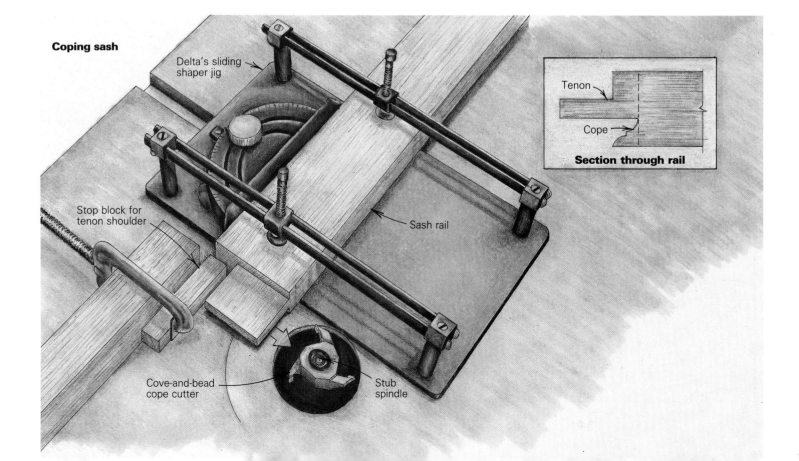

Coping sash

Delta's sliding shaper jig

Stop block for tenon shoulder

Cove-and-bead cope cutter

Stub spindle

Sash rail

Tenon

Cope

Section through rail

ta makes for their shapers (drawing, facing page). This jig combines a miter gauge and a small sliding table with integral hold down clamps. I coped the ends of the muntins before molding their edges so that I would have a flat surface bearing against the miter gauge. The subsequent edge-molding operation would remove the tearout left by the cope cutter. Sizing muntins individually lets me use up a lot of otherwise useless short and narrow scrap, as well as making it easier to account for all the necessary pieces before moving on to the next step.

To ensure perfect mating of cope and sticking, I set up both operations simultaneously on separate shapers. If I only had one shaper, my first step would have been to make an accurately fitting pair of prototypes by trial and error (one stuck, the other coped). These would have been used to test the various setups. The stiles and rails were beefy enough that I felt I could run them safely on the shaper without holddowns. For the muntins, I used Delta's nifty spring hold-downs, which held the small molding securely to the fence and table.

A wide straight cutter could have been used below the sticking cutter to do the rabbeting at the same time, but it would have meant making a heavier cut than my equipment is comfortable with. Instead, I used a narrow straight cutter along with the sticking cutter. This ploughed out part of the rabbet. I sliced off the remaining material on the tablesaw.

Assembly and installation—After all the shaper and saw cuts were complete, the sashes were ready to assemble. The nice thing about through-tenoning is that it allows me to start all the tenons in their respective mortises. Then all I have to do is brush a light coating of glue—resorcinol in this case—into the mortise from the opposite side and tighten the clamps. That saves a lot of frantic moments. I used clamping blocks that have a channel ploughed in one side. Centering this channel over the mortise before tightening the clamps allows the tenon to protrude slightly beyond the edge of the stile if necessary.

After laying the insulated-glass units into the sashes and shimming them into position, I was faced with the task of fitting all the moldings that would hold the glass in place. One of the rooms I worked in, for example, contained six sidelights, five casements and one 8-ft. pair of French doors. That translated into 112 panes of glass, which meant 448 moldings requiring 896 miters. There had to be a better way.

The slotted track and crank housing of the casement operator are simply screwed in place. The "crank molding" is the narrow stock notched to fit over the crank housing. It provides a surface for the screen to sit on. The solid-brass hinges have a small set screw in the barrel that prevents the hinge pins from being removed from the outside.

My shortcut was to cope wide stock on the shaper first, then mold the sticking on the edges and rip off individual precoped moldings. This trick saved many tedious hours hunched over a screaming miter box. Another time-saver I discovered was a pneumatic brad tacker. It came in mighty handy for the 2,688 fasteners.

Hanging the sash was no different from hanging doors. I used a router and a plywood template to cut the hinge mortises in the sash. On the new jambs that I made, I had routed mortises for the hinges before assembling and installing the frames.

Because the hinge barrels for each sash would be located on the outside, I wanted solid-brass butt hinges to resist corrosion. I also needed the security of a nonremovable pin. The ones I used were made by the Baldwin Hardware Corp. (841 E. Wyomissing Boulevard, Box 15048, Reading, Penna. 19612) and cost $50 a pair. The end caps on the barrel of the hinge unscrew, allowing removal of the hinge pin for easier installation of the sash. A tiny set screw on the indoor side of the hinge barrel secures the pin afterwards. The casement locks were simply screwed to the sash and their strike plates drilled and mortised like those for the lockset on a door.

I worked out the location for the casement operator by trial and error. If the crank was too close to the jamb, there wasn't enough room for a hand to turn it. And if the crank was too far from the jamb, the window wouldn't open very far. Once I figured out the best location, installation was straightforward—a slotted track was screwed to the inside face of the sash and the body of the casement operator was screwed to the window stool. I made a notched wooden piece, which I call a "crank molding," to fit over the casement operator (photo left). In addition to being notched for the body of the casement operator, it is rabbeted in the back to receive the retracted arm of the casement operator when the sash is closed. There's also a shallow rabbet in the top for the screen to sit in. I made all the cuts on the tablesaw, starting with the crossgrain notch for the operator.

Weatherstripping the windows was a prime concern. I chose a stainless-steel leaf-type material called Numetal, made by Macklanburg-Duncan (Box 25188, Oklahoma City, Okla. 73125). Each box of Numetal includes nails and about 30 ft. of weatherstripping. I nailed it to the side and head jambs along the sash rabbet. It is fairly easy to apply, invisible when the sash is closed and is the most durable of any system I've used.

On the bottom of the sash, I used a rubber sweep on the outside to seal the sash against the sill. Then I mounted a piece of compressible gasket-type weatherstripping on the stool for the sash to press against. I could have run it along the sides and head as well, but the material would have interfered with the fasteners on the strike side and might have gotten pinched by the closing sash on the hinge side. It wouldn't have looked very good, either.

For the screens, I built simple frames out of ¾-in. stock. I cut a rabbet on both sides—a shallow one on the inside for decoration and a deeper one on the outside for the screen and screen bead molding. I coped the rails into the stiles, which essentially creates a shallow bridle joint, glued them together and ran a long screw through the ends. I cut a narrow screen-retaining groove in the bottom of the rabbet on the screen side and used a screening tool (which looks like a dull pizza cutter) to press in the aluminum screen. I used a utility knife to trim the screen against the corner of the rabbet and then nailed on the screen bead molding. To hold the screens in place, I installed bullet catches between the stiles and jamb—one on each side, about halfway up the stiles. A delicate brass knob, installed on the face of the frame, makes the screens easy to pop out in the fall for a clearer view. □

Scott McBride is a carpenter in Irvington, New York, and a contributing editor of Fine Homebuilding.

Installing Fixed Glass Windows

Double-glazed units that don't move let in light and heat, but keep out drafts

by Dale McCormick

Windows have traditionally provided light and ventilation, but asking them to perform both functions isn't always wise, especially when many of us are installing large expanses of glass on our south walls. Windows that don't open can be sealed more tightly than those that do, and they cut down on the infiltration that often amounts to 40% of a house's heat loss. Ventilation can be handled by screened wall openings, carefully placed and built to be heavily insulated and tightly sealed during cold weather. Even if you like movable sash for ventilation, you can intersperse it with fixed panes.

Insulated glass—A single pane of glass is a lousy insulator, typically yielding an R-value of less than 1. Two spaced panes trap a layer of still air between them and can cut heat loss in half. Three panes create two air spaces and cut heat loss even further, but triple glazing isn't cost-effective in most areas, especially if you're planning to use some form of insulation inside the windows at night. The problem with double glazing has always been to eliminate condensation between the panes while at the same time sealing them to prevent convection from destroying the insulation value of still air. Although you can build your own double-pane insulating windows by installing separate sheets of glass in a wood frame, moisture will invariably migrate through the wood, and the resulting condensation looks bad and can lead to rotting frames. Weep holes aren't a good solution—you don't want air movement. I recommend buying ready-made double-glazed units.

Commercial insulated windows are built of two panes of glass separated in an aluminum channel. The channel contains silica gel, a des-

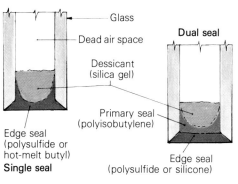

Factory-sealed insulating glass
Dual-seal units are more resistant to punctures or other damage that could destroy the insulating value of the air space. They generally cost more and come with longer guarantees.

sicant that absorbs moisture trapped between the panes of glass at the factory. The glass and channel are sealed with either one material or two (drawing, above). When only one seal is used, it is usually hot-melt butyl or one of the varieties of polysulfide. Such units are often guaranteed for five years. Better windows have two seals. The primary sealant, a moisture barrier, is polyisobutylene. The edge sealant is either polysulfide or silicone. Don't confuse any of these materials with the glazing sealant you or your contractor will have to apply when installing the glass. It's important for the factory edge sealant and your site-applied glazing sealant to be compatible—many aren't.

Jambs—A jamb is a frame that holds the window and is set within the rough opening in the side of the house. Fixed-pane windows can be built either with or without a jamb. Using one

means more work, but jambs are independent of the house's framing, and can be plumbed and leveled within the rough opening. If you don't use a jamb and attach your windows directly to the studs, the fate of the glass is wedded to the future of the frame. The key to a good marriage is dry wood and accurate framing. Generally, fixed windows with jambs look more finished, but you can also achieve an attractive effect without using a jamb, if you choose a handsome wood as the structural material for the window wall.

For large expanses of glass, it usually isn't practical to invest the time and material necessary to build a jamb for each pane. There are a number of ways to install double-glazed windows without jambs. One method we often use is shown in the drawing at the top of the facing page. The glass, attached to the outside of the studs, is held in place by the casing, which doubles as a stop. We set each glass unit to bear ⅜ in. on the studs, leaving a ¾-in. gap between units if we're framing with nominal 2x4s.

Manufacturers of insulating glass recommend that the units be installed with ⅛-in. interior and exterior face clearance between the glass and the stop. For this, we use butyl glazing tape all around, which functions both as a bed for the glass and as a dam to prevent the caulking from touching the edge seal. Butyl also reacts with the primary sealant, so keep their edges separated. This is easy if you use narrow tape.

We set the glass on two neoprene blocks measuring ½ in. by 4 in. by the thickness of the unit. These should be positioned in from the edge of the window by one-quarter the length of the glass panes. Last, we caulk with a waterproof material compatible with the edge seal-

Illustrations: Barbara Smolover

Tilt? While you are designing your house or addition, you have to decide whether to install your windows vertically or angle them to accept more solar radiation. I've stopped using slanted south-facing glass except in a few special situations: a space that will be used mostly to grow plants; a space that can be closed off from the house so that overheating in summer and night insulation in winter are not problems, or a retrofit of a narrow south porch, where slanting the glass from the roof eave to the ground will yield more room.

Angled glass creates what is basically a glass roof, inviting a multitude of problems: leakage, breakage,

An aluminum glazing system

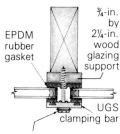

EPDM rubber gasket

¾-in. by 2¼-in. wood glazing support

UGS clamping bar

#14 hexhead screw 2½ in. long, 16 in. o.c., with weatherseal washer

overheating in summer and snow cover in winter. Also, in the summer, when you don't need it, many more Btus come through glass angled at 60° than through vertical windows—and vertical glass can be more easily shaded. On the other hand, there is a very small difference in performance between 60° glass and 90° glass during December and January, because the low winter sun is more nearly perpendicular to vertical windows then.

If you do choose to angle your glass, it's probably best to use a commercial-style aluminum glazing system with EPDM rubber gaskets (drawing, left). Aluminum won't expand or twist as wood does under the extreme conditions faced by a roof surface oriented south.

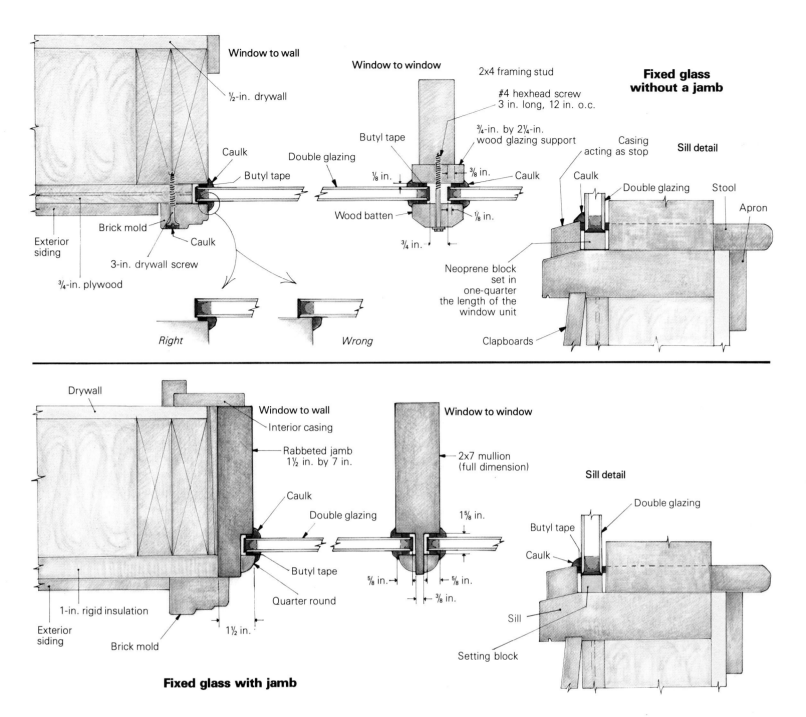

Window to wall

½-in. drywall

Caulk

Butyl tape

Double glazing

Brick mold

Caulk

3-in. drywall screw

Exterior siding

¾-in. plywood

Right   *Wrong*

Window to window

2x4 framing stud

#4 hexhead screw 3 in. long, 12 in. o.c.

Butyl tape

¾-in. by 2¼-in. wood glazing support

⅛ in.

⅜ in.

Caulk

Wood batten

⅛ in.

¾ in.

Neoprene block set in one-quarter the length of the window unit

Fixed glass without a jamb

Casing acting as stop

Sill detail

Caulk

Caulk

Double glazing

Stool

Apron

Clapboards

Drywall

Window to wall

Interior casing

Rabbeted jamb 1½ in. by 7 in.

Caulk

Double glazing

Butyl tape

Quarter round

1-in. rigid insulation

Exterior siding

Brick mold

1½ in.

Window to window

2x7 mullion (full dimension)

1⅝ in.

⅝ in.   ⅝ in.

⅜ in.

Sill detail

Double glazing

Butyl tape

Caulk

Sill

Setting block

Fixed glass with jamb

ant. As shown in the drawing, caulk should actually cover an opening, not just be run along a crack. Wood should be primed before assembly, because many caulks, including silicone, pull away from bare wood over time.

Last summer, we built an inexpensive greenhouse this way, using pressure-treated lumber to withstand condensation and moisture. The greenhouse looks appropriately plain and practical. We also used this method of glazing on a sunspace kitchen/family room in a summer house we were retrofitting for year-round use. Here we used lauan studs to frame the south wall, and lauan exterior stops to hold the glass. The result is elegant. It's not the method but the materials (and the care with which you use them) that determine how your window installation will look.

For relatively small areas of fixed glass, especially those that are integrated with operable windows, I think it's best to design fixed windows with a traditional jamb that can be made in the shop. Exterior casing that matches the

casing of movable sash will visually tie the two types of window together.

Jamb stock should be dry. Pine or Douglas fir is fine. Redwood is a good choice for greenhouses. Rip your stock to a width equal to the distance between the interior and exterior finish materials. You can rabbet your jambs, as shown above, to produce an integral stop. If you do, the rabbet depth should match the size of the stops that you're going to apply to the outside—usually ¾ in. The width of the rabbet should be at least 1 in. greater than the thickness of the glass, depending on the thickness of the glazing tape you'll be using. An alternative to the rabbeted jamb is to use 5/4 stock (which is really 1 in. thick) for the jamb and add 4/4 interior stops, which can also act as stops for insulated shutters.

Either way, it's best to build the window sill from 7/4 stock. As shown in the drawing, the sill is wider than the rest of the jamb, and should be designed to overhang the exterior siding by at least 1 in. The sill should be bev-

eled so that rain will run off, and it should have a drip kerf to prevent water from running in along its bottom surface. A groove under the sill to accept siding material makes for an extra-tight installation.

In the shop, rabbet the tops and bottoms of the side jambs to accept the head and sill, keeping everything square. Glue and nail them into place with 12d ring-shank nails. The finished jamb unit can then be plumbed and leveled in the rough opening with shims under the sill and along the side jambs. When you install the glass, use the same clearances and techniques you would in a jambless installation, and be sure to leave ⅛ in. between the inside edge of the butyl tape and the interior edge of the rabbet. Brick mold is a good exterior trim treatment, because it closely matches the trim on commercially manufactured windows. ☐

Dale McCormick is in charge of building services at Cornerstones, an owner-builder school in Brunswick, Maine.

Designing and Building Leak-Free Sloped Glazing

Error-free projects demand careful detailing

by Fred Unger

Water has an uncanny ability to work its way into the most unexpected places, while the sun can quickly drive all life from an improperly designed sunroom. That's why sloped-glazing units—such as those in sunrooms, conservatories and large skylights—are among the most complex parts of the building envelope. They're also among the hardest to build correctly.

But with the right materials, careful attention to detail and a sense of how water behaves, well-functioning skylights and sunrooms are well within reach of most knowledgeable builders. Building a high-quality unit isn't cheap, but it will cost less over the long haul than endless callbacks or having to replace an entire structure later on. I tell sunroom clients that, while I can build a conventional small addition for $100 to $120 per square foot, they should plan to spend at least $160 per square foot for a comfortable, well-built sunroom with overhead glass.

Years of trial, error and refinement have passed since I worked on my first sloped-glazing project in 1974. I've learned that the secret to controlling water problems is a dry glazing system that depends not on caulks or sealants, but on gravity and physics. Before going into the details of the system, I'll review some general principles that apply to all wood-frame sloped glazing.

Custom or manufactured?—We custom-build many of our sloped-glazing units on site or in our shop. But why custom-build when there are plenty of manufactured products on the market? Part of the answer is that some glazing systems available to custom builders are superior to those found in most manufactured units (I can't understand why manufacturers have been so slow to use them). Although we do install some manufactured systems, we only work with companies that will customize their units to our specifications and with units on which we can install our preferred glazing system. To save time in the field and enhance quality control, we also pre-

glaze some skylights in our shop, truck them to the site and hoist them into place with a crane. Most manufactured systems are designed to be site-assembled; thus, they aren't engineered to withstand the added stresses of transport and hoisting.

The other part of the answer concerns design. Architects and clients can be quite creative in their designs (for example, we recently completed an African mahogany pyramid skylight with heat-mirror glass), but there are limits as to how far manufacturers will go to customize their products. So far, we haven't found a manufacturer that meets all of our demands for design and detailing.

On architect-designed projects, we try to get involved early in the design process. By sharing our expertise, we can help specify the framing materials, glass and sealants. We also suggest details that will expedite the construction process and save the client money. Our goal is a durable, efficient, cost-effective structure that meets the architect's and the client's design goals. Offering such help has gotten us specified into projects without competition.

Choosing a frame—The most critical design factor we must deal with is movement. On a sunny winter day, the components of a south- or west-facing skylight or sunroom may be exposed to some extreme temperature changes. The differing rates of expansion and contraction of wood, metal, glass and rubber can lead to buckled flashings, torn caulk joints, failed water seals and even broken glass. The movement is even more pronounced in poorly designed, wood-framed spaces that house plants, a pool or a hot tub. I've seen sunrooms built with construction-grade lumber in which the frames had twisted or warped well over an inch.

Because movement is so potentially damaging, a dimensionally stable frame is a must (photo right). The use of unstable wood can lead to major problems even if all other aspects of the project are executed perfectly. We prefer to work with clear, kiln-dried, vertical-grain stock. The most dimensionally stable

Materials count. A skylight or sunroom must withstand wind, rain and sun, so even a simple project demands a dimensionally stable frame. The best domestic woods are clear, vertical-grain redwood and cedar (photo below). Outside, hips, valleys and angles can be covered with custom aluminum caps that can be fabricated at a sheet-metal shop (photo right). They're held in place by exposed stainless-steel screws with gasketed washers.

species we've found are redwood and cedar, although we've also had good luck with some species of mahogany. Good wood can be quite expensive. Luckily, however, laminated wood is also an excellent choice: It's less expensive and more environmentally friendly than solid stock cut from old-growth trees. It's also more stable. We never use solid sawn hem-fir, white pine, yellow pine or oak except as part of a composite member.

Regardless of the species, the best glazing systems we've found require the surface on which two pieces of insulating glass come together to be about 3 in. wide. The bearing area can consist of solid wood or of glued 2x or 1x stock. If these pieces will be exposed to high humidity, a two-part resorcinol or other waterproof glue should be used.

Precision and flexibility—We try to design our frames to meet two seemingly contradictory demands: precision and flexibility. The joinery on most of our projects is complicated, especially when hips and valleys are included. Because virtually every cut ends up as exposed finish work, our frames have to be precise. Structural connections demand careful planning and intelligent detailing. Once ordered, expensive insulating glass can't be trimmed to fit improperly framed openings. Because of this, we must design and build our frames to close tolerances.

We usually join our framing components with blind screws, dowels, biscuits or other hidden fasteners. We've also used exposed brass screws. Some areas, such as structural ledger boards, can be hidden. We attach them with nails or screws and then hide them with trim. Our choice of fasteners depends on the expected humidity level and the wood species we're using. In a high-humidity environment, the natural extractives in cedar and redwood will corrode even galvanized fasteners. We tend to use a lot of stainless-steel screws and nails.

Despite the fact that our frames must double as finish work, they also have to perform in the real world of construction. We can't expect framers, masons and foundation crews to work to $\frac{1}{16}$-in. tolerances in three dimensions. We compensate by designing shim spaces into the system and then use trim to span the gaps.

The system—Many glazing systems rely on caulks and sealants to prevent water problems. But we've found sealant-dependent systems to have some real disadvantages. While some of them will stop leaks if installed under near-ideal conditions, most won't control condensation. These systems also make glass replacement difficult. I've spent hours with scrapers and solvents, removing broken glass that had been caulked in place. But it wasn't until a client's dog stepped on a gob of urethane caulk and tracked it across a $5,000 rug that I became determined to minimize our use of these products.

The dry glazing systems we now use employ rafter and purlin baseplates with internal gutters, EPDM rubber gaskets, insulating glass and a gasketed aluminum pressure cap (drawings right). The aluminum cap holds the glass in place. EPDM gasketing or closed-cell foam tape serves as the primary water seal. A removable trim cap hides the screws, which are installed with gasketed washers. On purlins, we use flat bar stock over closed-cell glazing tape because the resulting low profile prevents water damming. The glass sits on raised pads that prevent any water in the guttering system from puddling against the window's edge sealant. Such puddling is a leading cause of seal failure and fogging within insulated glass. To keep the glass from sliding and to keep adjacent pieces of glass from shearing relative to each other, we rest its bottom edge on rubber setting blocks and aluminum setting-block supports. These blocks support the glass and keep top and bottom lites from shearing relative to each other. The supports are positioned one quarter of the way in from the bottom corners of the glass.

But the guts of this system are the internal gutters in the baseplates. Good systems are designed to direct water from the purlin gutters to the rafter gutters to weep holes at the eaves. In effect, the gutters serve as a safeguard to the primary water seal. As one old-time glazer told me, "If you don't want water problems, you've got to design the system to leak." The dry system also permits quick, low-cost replacement of broken glass.

We use these systems because we consider them to be the most foolproof. Even though we'd like to pretend that our designs, our installers and our materials are perfect, they're not. For best results, workers should understand how water moves, along with why and how the system sheds water. Fortunately, however, the details we've developed will work even if installed on a Monday morning. On one job, we set the glass and then got a solid week of rain before we could cap the system off. We didn't get a single leak.

Baseplates can be aluminum with EPDM gaskets, or solid EPDM rubber (photo page 46). The best we've found are available from Abundant Energy, Inc. (P. O. Box 307, County Rt. 1, Pine Island, N. Y. 10969; 800-426-4859) and U. S. Sky (2907 Agua Frio, Santa Fe, N. M., 87501, 800-323-5017). The aluminum systems have been around for several years and many now include thermal breaks. Aluminum baseplates hold screws for the pressure caps more securely than wood systems do, and ensure a flat surface for glazing. But they take more time to install than the rubber ones and are hard to work around hips and valleys without adding cumbersome-looking details. If they don't include a built-in thermal break, aluminum baseplates themselves can cause minor condensation and interior frosting. Even thermally broken aluminum systems must be carefully detailed to prevent thermal conduction problems.

The EPDM baseplate, on the other hand, installs quickly and easily (it cuts with shears or

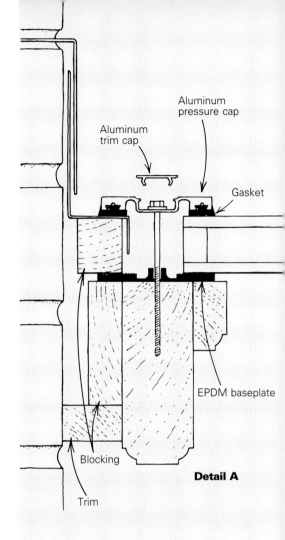

Aluminum trim cap — Aluminum pressure cap — Gasket — EPDM baseplate — Blocking — Trim

Detail A

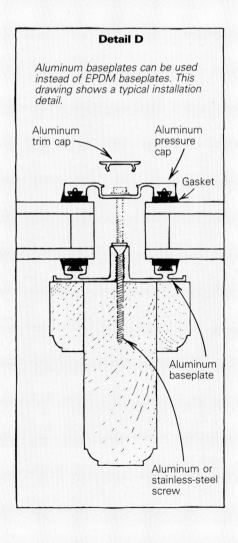

Detail D

Aluminum baseplates can be used instead of EPDM baseplates. This drawing shows a typical installation detail.

Aluminum trim cap — Aluminum pressure cap — Gasket — Aluminum baseplate — Aluminum or stainless-steel screw

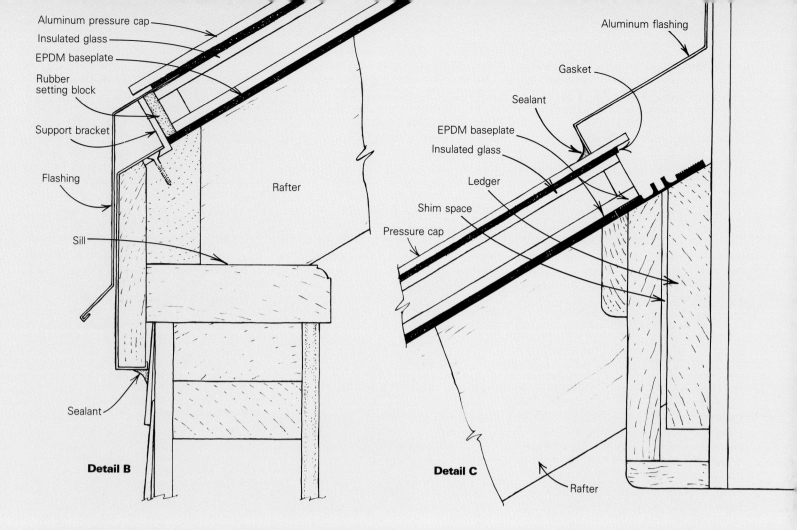

Aluminum pressure cap
Insulated glass
EPDM baseplate
Rubber setting block
Support bracket
Flashing
Sill
Sealant

Rafter

Detail B

Aluminum flashing
Gasket
Sealant
EPDM baseplate
Insulated glass
Ledger
Shim space
Pressure cap

Detail C

Rafter

Shedding water. *A dry glazing system is designed around a system of aluminum or EPDM rubber baseplates with internal gutters. By directing any leaks or internal condensation to weep holes at the eaves, the gutters serve as a backup to the primary water seal. The system also includes EPDM rubber gaskets, insulating glass and an extruded-aluminum pressure cap. The aluminum pressure cap holds the glass in place. Raised pads beneath the glass prevent any water in the gutters from puddling against the window's edge sealant. The bottom edge of the glass rests against setting blocks and setting block supports. In addition to shedding water very effectively, the system also permits quick, low-cost glass replacement.*

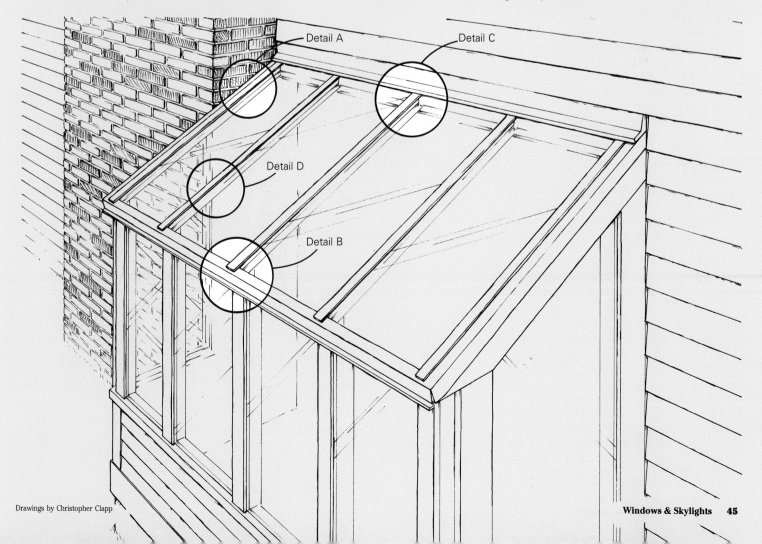

Detail A
Detail C
Detail D
Detail B

Drawings by Christopher Clapp

a utility knife), bends around hips and valleys and isn't prone to condensation or frost problems. Unfortunately, the EPDM baseplates that have been available don't have condensation gutters and cannot be easily installed to direct water from a horizontal purlin to a vertical rafter. They should only be installed on systems with one lite of glass vertically. Even then, the glass has to be carefully detailed.

Abundant Energy is about to introduce an EPDM baseplate that incorporates cascading internal gutters. It's fast and flexible to install and can be cut with a utility knife, yet it costs less than the aluminum system. It also has the fastest-to-install setting block support of any system I've found on the market, isolates water from any penetrations through the baseplate and eliminates any concern about thermal conduction.

Metals and flashing—Even with rubber baseplates, metals remain a crucial part of any sloped-glazing project. Baseplate gutters aren't intended to handle rivers of water, so it's important to have good exterior cap and flashing details. We clad the exterior sloped portion of all our projects with an aluminum cap (photo, page 43). Abundant Energy and U. S. Sky have excellent caps, as do most manufacturers of sunroom kits and curtain-wall glazing systems. Most of these caps are designed for use where two lites of glass meet on a flat plane. For hips, valleys and angles, we have a sheet-metal shop fabricate custom caps from heavy aluminum flat stock with an anodized or baked-enamel finish. These must be fastened using exposed stainless-steel screws with gasketed washers.

Occasionally, an architect will insist on copper or lead-coated copper flashings. But copper and aluminum are galvanically incompatible; using them together raises the possibility of corrosion (for an explanation of galvanic corrosion see *FHB* #62, pp. 64-67). When we have to put galvanically incompatible metals in close proximity, we make sure to isolate them with wide, closed-cell foam tape or with EPDM gaskets.

Metals expand and contract greatly with changes in temperature. Before screwing or nailing through caps or flashings, we always predrill or punch over-sized pilot holes. To protect against galvanic reactions, we use stainless-steel fasteners or fasteners of the same metal as the flashings. Prepainted aluminum flashings also can't be soldered. Where two or more lengths of flashing are required for a single run, we leave a ⅛-in. gap between them and then install a spline below the gap. When the edge of the flashing is bent over and splined, it locks the two runs together.

Sealants—Despite our best efforts to minimize them, we still use lots of sealants (for more on caulks and sealants see *FHB* #61, pp. 36-42). To prevent leaks, we always run a bead of silicone along the top edge of our purlin caps and eaves flashings, then extend it a couple of inches up the edge of the adjacent rafter caps. We also use sealants at butt joints in flashings. We use sealants that will permit high levels of joint movement; this is especially important when transporting preglazed skylights over bumpy roads.

At one time, we tried using only neutral-cure silicones. They're easily gunned, permitting a nice, clean bead, and they were the only silicones I could find that showed any tenacity in sticking to wood. Most silicones also adhere well in glass-to-metal connections. But we found that regardless of how well they adhered when first applied, the silicones would eventually begin to pull away from wood. We now use urethane sealants on all wood and masonry joints. The urethanes are absolutely remarkable in their adherence, though they're not as easily gunned or tooled as the silicones. We still use silicones for glass-to-metal connections (such as the leading edge of the eave flashing) or for metal-to-metal connections (such as the intersection between two flashings).

Pros and cons. **Baseplates can be aluminum or EPDM rubber. Aluminum provides a stable bearing surface for glazing but can be time-consuming to install, and it is subject to interior frosting. The EPDM doesn't frost up but doesn't have a condensation gutter. Photo by Susan Kahn.**

One thing to watch for when choosing sealants is potential chemical incompatibility between the field-applied products and the insulating glass-edge sealants. The chemical reactions between two incompatible sealants can lead to the breakdown and failure of one or the other. Several builders, glass companies and window manufacturers have been ruined by lawsuits that resulted from chemical-related seal failures. Your insulating-glass manufacturer should be able to provide sealant-compatibility test results upon request.

Caulks and sealants should only be applied to clean, dry surfaces. Before caulking, we clean adjacent surfaces with glass cleaner. To cut any oils or residual films, we usually wipe the joint with a rag dampened slightly with a solvent called xylol. This is also the best solvent we've found for removing silicone—which is why we take care to keep it away from the window edge sealants.

Choosing the glass—Insulating glass consists of two or more lites of glass separated by an air- or gas-filled space. A hollow aluminum spacer is joined to both laminations with a moisture-proof edge sealant. The aluminum spacer is filled with a dessicant that keeps the air between lites dry.

When ordering glass, be sure to check its warranty, code-compliance and thermal and solar characteristics. I'll touch on the first two here; the third is a subject for a future article. The glass used in a sloped-glazing unit should be guaranteed for use on a slope (most glass isn't). Putting glass on a slope adds differential stresses between the inboard and outboard lites that aren't present in vertical glazing. Most sloped-glazing units have a dual-edge sealant—a silicone structural sealant and a moisture seal consisting of some other material. A few companies are switching to a new single-seal urethane. If you're using urethane sealed units, it's important to ensure that all edges of the glass are capped. The sealant will break down if it is exposed to direct sunlight.

The second major consideration with sloped glass concerns safety and codes. Most codes require that sloped glass on commercial buildings include an outboard layer of tempered glass and an inboard layer of laminated safety glass. The tempered glass is strong enough for someone to stand on and usually won't break unless tapped on an exposed edge or hit with a sharp object. The plastic layer within laminated safety glass is strong enough to hold any tempered glass that does break, along with whatever broke it. These safety standards make sense for residential work, too; in fact, several states have incorporated the standards into their residential building codes. □

Fred Unger is the owner of Heartwood Building Specialties in Berkeley, Mass. He designs and builds custom additions, skylights and sunspaces. Photos by Charles Wardell, except where noted.

Screen-Porch Windows

Some site-built alternatives to manufactured window units

by Douglass Ferrell

These unobtrusive accordion windows fold together against the wall, but can be quickly drawn across the screened opening if a storm blows up.

Some of my best childhood memories took place on the old screen porch at my family's lake cottage. Out there around a big wooden table, we used to play cards and board games late into the night. Moths, mosquitoes and other bugs droned beyond the screens and across the dark lake. On rainy days we assembled jigsaw puzzles out on the porch and made napkin rings and other trinkets out of birch bark.

Ever since then, I have been interested in screen porches and what makes a good one. For me the key elements include wood framing, large expanses of screen, low window sills and a long wall facing the view, which might be a body of water, treetops seen from a rise or simply an expanse of green lawns.

The screen porches favored in the Sun Belt, with concrete-slab floors, aluminum framing and floor-to-ceiling screens, have the advantage of not requiring operable windows over the screens to keep out the weather. But these porches aren't suited to northern climates. The alumi-

num and concrete are too cold, and full-length screens forfeit an important sense of shelter.

Wooden screen porches, on the other hand, need some type of operable windows to protect both the structure and its furnishings from driving rain and drifting snow. Although the design and construction of the rest of the porch may be straightforward, the windows can be a challenge.

On any porch, a series of factory-made windows or glass doors with all their jamb and sash components cluttering up the view just can't

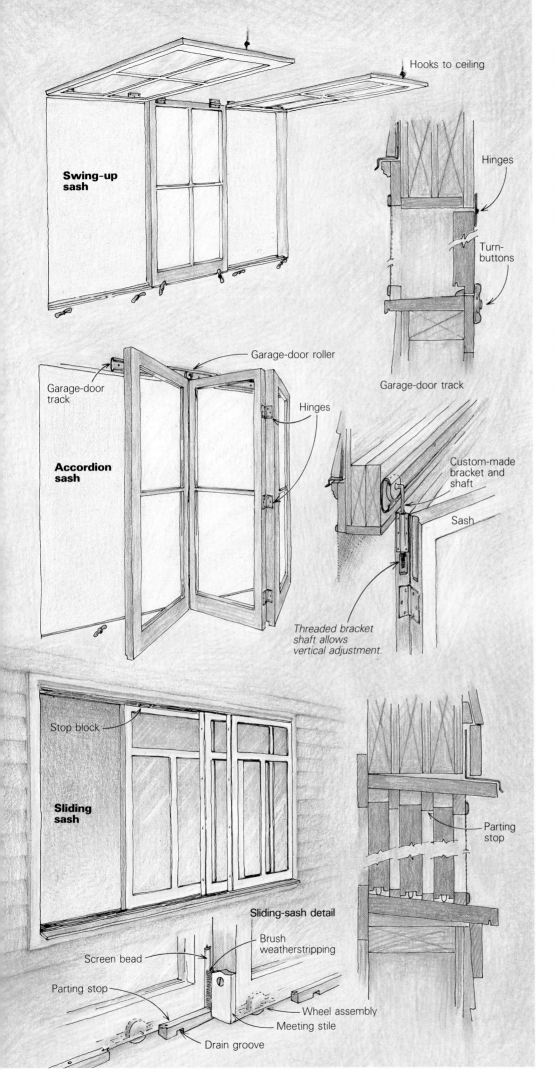

Swing-up sash

Hooks to ceiling

Hinges

Turn-buttons

Garage-door track

Accordion sash

Garage-door roller

Garage-door track

Hinges

Custom-made bracket and shaft

Sash

Threaded bracket shaft allows vertical adjustment.

Sliding sash

Stop block

Parting stop

Sliding-sash detail

Screen bead

Parting stop

Brush weatherstripping

Wheel assembly

Meeting stile

Drain groove

provide the airy feel of a good screen porch. In the ideal window system, the sash should store out of the way, be easy to operate and shouldn't interfere with furniture.

I will describe here three types of porch windows. One is an old standby. The other two I've made and installed with the help of my inventive uncle, Dave Stewart.

All my window systems use ready-made sash either in stock sizes from a lumberyard or as custom-made units from a millwork shop (though you could certainly make your own, if you're so inclined). These systems are also based on the assumption that the porches won't be heated full-time. Porches designed for more than occasional cold-weather use should probably incorporate factory-made units because site-built windows are difficult to weatherstrip effectively.

Swing-up sash—Swing-up windows (drawing, top left), hinged to the head jamb, are traditional favorites for porches and basements. They aren't expensive, and when open, look appealing in an old-fashioned way. Installing them is fast and easy, particularly if the finished openings are precise and square. If they're not, you will have to use sash slightly oversize and trim them one at a time to fit. To install the sash, just space it about ¼ in. above the sill and screw hinges to the header and the top rail of the sash. Hooks and screw eyes hold the sash open, and turnbuttons screwed to the sill hold them closed.

A lot of older cottages in the lake country of Wisconsin and Minnesota use this window system. But when a summer thunderstorm blows up, you have to rush around and move chairs, lamps and maybe a stray knitting bag in order to swing the windows down and keep the rain out. High sills and low furniture minimize the problem, but of course that cuts into the view.

Accordion sash—Many years ago my grandfather devised an improvement on the old swing-up sash system. His system consists of sash, hinged to fold up like an accordion, that slide along an overhead track made out of garage-door hardware (photo previous page). The dimensions of the finished opening are not as critical for this system because the top of the sash butts against the side of the head jamb, and the custom hanger bracket allows a vertical adjustment (drawing, middle left). This system is well suited to retrofits and structures that are likely to shift a bit over the years. On my installation I had the hanger brackets made by a small machine shop, but I suspect you could make do with one leaf from a large butt hinge.

To install the system, just fasten the sash together with butt hinges and sit them on ¼-in. spacers on the sill. Slide the hanger rail onto the rollers, pull it up snug, and screw it to the header. If the window doesn't operate smoothly after you remove the spacers, it's easy to adjust the hanger brackets in place by tightening or loosening the nut at the bottom of the roller shaft.

The units I built run wall to wall, as do the tracks. Once opened and folded, the windows lie flat against the sidewall, where they are held in place by a wooden latch mounted on the wall above them (photo next page). When the win-

Drawings: Christopher Clapp; Photos: Douglass Ferrell

dow is closed, turnbuttons along the sides and bottom of the assembly secure the sash.

These windows are a little cumbersome to operate, partly because a large window is fairly heavy, although well-fitting windows up to about 14 ft. wide work fine after a little practice. Rolling one window unit toward each corner allows a full-width opening in a 28-ft. or 30-ft. wall—big enough for most houses. Windows with low sills can theoretically clear furniture placed along the wall by pulling and folding the sash from the corner while the unfolded sash roll toward you, but of course there is some tendency for the sash to fold up in the middle of the unit while you're pulling from the corner.

Sliding sash—My inspiration for sliding windows came from an old cottage whose windows had a groove in the bottom of the sash that slid on top of a rail on the sill. While this approach might work fine with smaller (and very straight) sash, swelling wood in summer humidity jammed nearly every window in the old cottage.

With the help of my brother and uncle, I devised a system of sliding windows that works much better and minimizes problems with low sills. In this system, the sash roll on wheels let into their bottom rails and are guided by parting stops screwed to the sill and head jamb (bottom drawing, facing page). The wheels were supplied installed in the sash by a millwork shop, but similar ones are available from large hardware outlets. I used nylon wheels instead of steel because they are quiet and won't rust.

This window system looks straightforward, but the installation can be time-consuming. The window-jamb dimensions are critical with this system since each sash slides along the entire opening and can't be trimmed to fit in specific places. On the installation I did, I wanted to minimize casing, so I framed the rough opening carefully, planning to fasten the sills and jambs directly to the framing. But I had to shim the jambs in some places anyway. Of course this meant that the sash for these openings had to be cut down. One off-size header, a slightly bowed post and a few other reminders that wood is a natural material convinced me that

next time I would shim a finish frame into the rough opening just like a door jamb. This would ensure a precise opening for the sash to roll in without time-consuming trimming and adjusting.

The sash are held in place with stops and slide to one end of the opening. I used clear redwood for all the stops, jambs and sills because it is durable, easy to work, and resists warping. The bottom stops have a drain slot routed across their bottoms every 18 in. so they don't trap water on the sloping sill. They are screwed to the 6/4 redwood sills with brass screws. The sills themselves are screwed to the rough opening from below to minimize holes and hardware in the top of the finish sill.

On one 18-ft. wall where I used these windows, four 1⅜-in. wide sash roll on the sill, and I was concerned that the required sill would look awfully wide. Consequently I used narrow stops, which I wouldn't do again. I used ½-in. by ½-in. stops on the bottom and ½-in. by ¾-in. stops on the top. A better size would have been ¾ in. by ½ in. on the bottom and ¾ in. by 1¼ in. on the top. These larger stops would leave enough room between the head jamb and the top of the sash to pick the sash up and out of the track for cleaning and repair. Also, the stiffer track would better support the heavy sash.

The sill width I ended up with, 7¼ in., doesn't look cumbersome, given the big opening. Another inch or so in width, with a correspondingly wider stop, would have been better.

The sliding sash are sealed along their length with brush weatherstripping (bottom drawing, facing page). This operation went very smoothly. I worked from inside with the sash mounted in their tracks in the closed position. At each place where two sash came together, I fastened a strip of screen bead to the outer sash. Then I ripped 3/4 stock a little narrower than the sash for use as meeting stiles.

After cutting the meeting stiles to fit between the stops, I attached a piece of brush weatherstripping along their edges. This weatherstripping comes in rolls and has an adhesive backing (I added a few staples for insurance). The stiles were then each held in place against the screen bead, scribed and planed to fit flush with the face of the inner sash. With such big windows, this step was necessary because the stiles were not all perfectly straight, and the meeting stiles needed to fit snugly against the full length of the screen bead.

Once all the fitting was done, I screwed the meeting stiles to the sash with brass screws. If the sash warp in the future and ruin the fit of the weatherstripping, the meeting stiles can be unscrewed, adjusted and refastened. To make it easy to line up these weatherstripped joints when closing the windows, I screwed a stop block to the head jamb to locate each window.

Because wood does shrink and swell, the tolerances are tricky in any system where wood slides in wood. Of course a warped rail is also a source of trouble. When I visited the house two years after building my sliding windows, I could still operate each sash with one hand and quickly open or close a whole bank of windows. □

Accordion windows are held in place against the wall by a wooden latch.

Douglass Ferrell lives in Trout Creek, Mont.

Simple Joinery for Custom Windows

Time-saving lap joints can save money, too

by David Frane

Few builders enjoy midstream design changes, but most of us have learned to deal with them. There are times, though, when a small change can cause a big problem. An example of this came while we were building a large home outside of Boston, Massachusetts. The roofer had finished about half of the red cedar shingle roof when the architects literally threw us a curve: The client had decided to dress up the front facade with an eyebrow window installed high above the main entry (photo right).

This design change created two problems. First, our window supplier wanted too much money (about $3,000) to custom-build the unit; second, they couldn't deliver it for three months. We didn't want to frame the dormer without the window—the fits would be too exacting and mistakes too expensive to fix—so we decided to build the eyebrow ourselves.

After studying the architects' rough elevations, I told job supervisor Harry Irwin that we could build the unit for about half the quoted price—*if* the client would accept some unconventional and slightly archaic construction details (we actually did it for about one-quarter of the original quote). And we could have the window ready to install in 10 days. The architects and the client, pleased that we'd found an affordable alternative, quickly gave us the go-ahead.

Looking out from an empty attic, the unit (top drawing, facing page) would feature a four-lite hopper (a bottom-hinged window that opens inward). On either side would be two fixed sashes, and the entire assembly would be surmounted by a curved top jamb screwed to a sill beveled to improve drainage. A curved casing applied to the top jamb would serve as a crown molding over which the roof shingles would extend. The operable window was overkill; the owners previously had lived in old houses with hot attics, so they wanted to have plenty of ventilation up there this time. As it turned out, the hopper is rarely opened.

My plan was to build the individual window sashes first, screw them to the sill and then laminate the jamb around them. I could then apply the window stop and casing and haul the unglazed unit to the roof to be framed into place. After the roof was finished, someone would have to go back up and glaze the sash.

Simplified joinery—To save time, we decided to use old-fashioned glazing compound instead of curved wooden stops to hold the glass in

An unexpected curve. Lap joinery, used by the author to build this eyebrow window, offers an alternative to cope-and-stick joinery. Photo by Charles Wardell.

place. The window would be 35 ft. off the ground, so nobody would notice the substitution. And rather than using standard cope-and-stick joinery, I designed the sashes to incorporate lap joints.

Using lap joints probably saved a lot of time. Also, the multiple shaper setups and the jigging required for coping and sticking curved stock would have busted our budget. Though lap joinery is unconventional for a window, it's as strong as the conventional method. Besides, from the ground it would be indistinguishable from the joinery on the house's other windows.

Laying out the curves—Lap joints notwithstanding, making an eyebrow window remains technically demanding. The unit must be symmetrical and its curves fair—they must flow smoothly into one another with no visually jarring transitions. The overall shape of this unit

would be formed by three tangentially intersecting arcs having identical radii (bottom left drawing, facing page). Two concave curves—the ends of the window—would meet the ends of a single convex curve—the window's center arc. The window would measure 12 ft. end to end and 2 ft. from the sill to the apex of the center arc.

We began by drawing a full-scale pattern of the window's perimeter (representing the entire unit with the casing installed) on ¼-in. lauan plywood. Jim Garry, a member of my crew who has lots of shop experience, laid out the curves with a long trammel. It took some work to find a combination of centerpoint locations that produced the right shape with the proper dimensions.

Next came my turn. I got the profiles of the three sashes by drawing a series of lines parallel to the inside of Garry's pattern. These lines represented the lower edges of the casing, head jamb and window stop, as well as the curved top

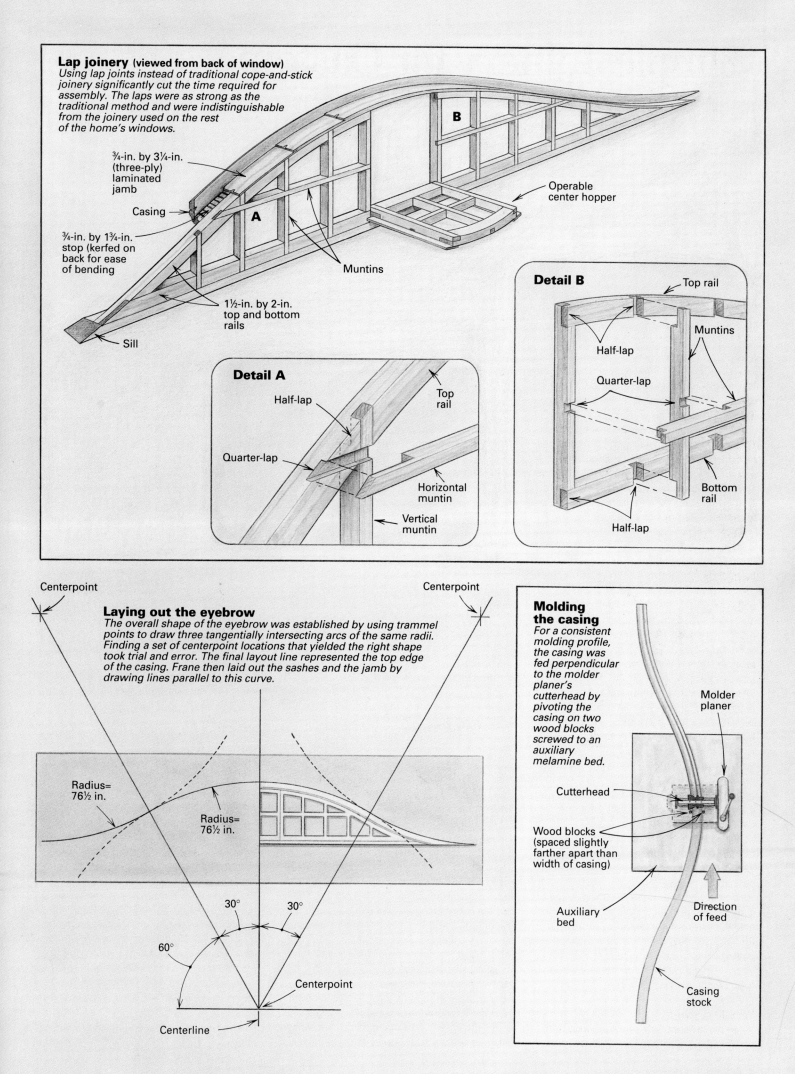

Lap joinery (viewed from back of window)
Using lap joints instead of traditional cope-and-stick joinery significantly cut the time required for assembly. The laps were as strong as the traditional method and were indistinguishable from the joinery used on the rest of the home's windows.

¾-in. by 3¼-in. (three-ply) laminated jamb

Casing

¾-in. by 1¾-in. stop (kerfed on back for ease of bending

1½-in. by 2-in. top and bottom rails

Sill

Muntins

B

A

Operable center hopper

Detail A

Half-lap

Quarter-lap

Top rail

Horizontal muntin

Vertical muntin

Detail B

Top rail

Muntins

Half-lap

Quarter-lap

Bottom rail

Half-lap

Centerpoint

Centerpoint

Laying out the eyebrow
The overall shape of the eyebrow was established by using trammel points to draw three tangentially intersecting arcs of the same radii. Finding a set of centerpoint locations that yielded the right shape took trial and error. The final layout line represented the top edge of the casing. Frane then laid out the sashes and the jamb by drawing lines parallel to this curve.

Radius= 76½ in.

Radius= 76½ in.

30°   30°

60°

Centerpoint

Centerline

Molding the casing
For a consistent molding profile, the casing was fed perpendicular to the molder planer's cutterhead by pivoting the casing on two wood blocks screwed to an auxiliary melamine bed.

Molder planer

Cutterhead

Wood blocks (spaced slightly farther apart than width of casing)

Auxiliary bed

Direction of feed

Casing stock

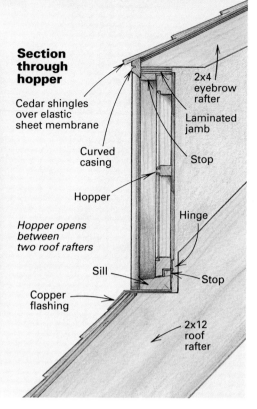

Section through hopper

Cedar shingles over elastic sheet membrane

Curved casing

Hopper

Hopper opens between two roof rafters

Copper flashing

Sill

2x4 eyebrow rafter

Laminated jamb

Stop

Hinge

Stop

2x12 roof rafter

Simplified framing, too. The eyebrow dormer was framed with 2x4 rafters, and the curved jamb was used as a header. Because the window looked out from an unfinished attic, the roof rafters behind it could be left intact. A pine batten is being used here to lay out a fair curve on the main roof.

rails of the sashes. I generated each line by making a series of pencil marks the proper distance down from Garry's curve, bending a thin wooden batten so that one of its edges touched all the marks, then connecting the marks with a pencil line scribed along the batten. At the bottom of the pattern, I drew straight lines for the sill and the lower sash rails, then drew in the muntins.

Making curved sashes—The first step in making the curved top rails of the three sashes was to transfer our curves from the lauan pattern to the solid white-pine sash stock. To do this, I borrowed an old boat-building trick (drawing below). I laid 5d box nails flat on the pattern stock with their heads on the curved line I wanted to transfer. I tapped the nail heads into the lauan so that they wouldn't move, then carefully lowered the sash blanks onto the pattern and tapped down on them to make an imprint of the nail heads on the blank. Then I drew lines through these marks, roughed out the blanks on the bandsaw and smoothed them with a spokeshave.

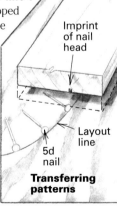

Imprint of nail head

Layout line

5d nail

Transferring patterns

Assembling the blanks into sash frames was straightforward. I simply laid the blanks on the pattern, ticked off the intersections of the stiles and the rails on the sashes and cut my half-laps with a dado head on the radial-arm saw. On curving cuts, I roughed out the rabbets with the dado head and finished them with a chisel. Routing would have been more precise, but it also would have required more setup time and would have given me a better job than I needed in this case.

I glued all the pieces together with an epoxy adhesive. For this job I used Chem-Tech's T-88 (P. O. Box 70148, Seattle, Wash. 98107; 206-783-2243). This product is as thick as honey and harder to use than many other epoxies. But we were working in an unheated shop and T-88 is the only epoxy I know of that will set at temperatures down to 35° F.

I glued the half-lapped stiles and rails together on top of the full-size pattern. Clear plastic sheeting placed between the sash and the pattern kept the two from being glued together. The sashes were to be paint grade, so instead of clamping the joints, I simply drove long drywall screws with washers through the joints and the lauan pattern and into the underlying wooden table.

When the epoxy set, I removed the screws and cleaned up the excess epoxy (the screw holes would be puttied later). Then I transferred the muntin locations from the pattern to the new sash frames. My preferred tool for dadoing the half-laps would have been the radial-arm saw, but our saw didn't have enough throw to reach across the sashes. Instead, I made multiple cuts with a Porter-Cable trim saw, then cleaned out the dadoes with a chisel. Dropping the vertical muntin bar stock into these dadoes, I then marked the locations where the horizontal muntins would intersect vertical muntins, stiles and rails. Because the horizontal muntin at each end of the window would intersect a half-lap on the top rail (detail A, previous page), I decided to install all of the horizontal muntins using quarter-laps instead of half-laps. I cut these quarter-laps by running the entire sash frame through the table saw, keeping the bottom of the sill against the rip fence. After testing for fit, I glued the muntins to the sash frames. Using a rabbeting bit in a laminate trimmer, I then cut rabbets on the exterior of the window to accept the glazing and the putty. The rabbets were squared with a chisel. On the interior, I used the router to cut an ovolo profile around each lite opening (hardly anyone

will ever see this, but it made me feel better). Now it was time to make the sill and the jamb.

Beveled sill, curved jamb—Making the sill and the jamb was the simplest part of the job. The sill is a single piece of 1½-in. thick pine ripped and beveled on the table saw. Because the sill would butt against the head jamb at both ends, I marked the end cuts of the sill by tacking the sashes to the sill and projecting the curvature of the jamb across the edges of the sill. Then I removed the sashes, cut the sill on the bandsaw, reattached the sashes and turned my attention to the jamb.

I laminated the jamb out of three 4-in. wide by ¼-in. thick strips of pine. Instead of building a separate laminating form, though, I used the tops of the sashes themselves. With the help of Don Pascucci (who would frame the unit onto the roof), I nailed the first lamination to the tops of the fixed sashes, then epoxied and nailed the successive laminations over it. After the glue set, I planed the rough edges of the jamb with a portable power plane, then trimmed its ends flush with the bottom of the sill.

Finally, I nailed a curved 1x1 pine stop to the top jamb. Because the stop would show a mere ¼-in. reveal beneath the top casing, I bent the stop by cutting a series of cross-grain ⁷⁄₁₆-in. deep kerfs in its backside. Epoxy secured it to the jamb, and putty filled the kerfs.

Milling reverse curves—Garry made the casing from 5/4 stock. After transferring the curves from the pattern to the stock, he cut his pieces and joined them into a single casing blank using long scarf joints and epoxy. The pitch of the scarf must be 1-in-12 or less; otherwise you're just gluing end grain, and the joint won't hold.

Of course, the main challenge posed by the casing was that, unlike the rest of the window, there was no way to escape having to mold a profile on it. We had run plenty of curved stock

Where to buy eyebrow windows

The following companies offer either stock or custom windows that can be used for eyebrows. This is not a complete list, however. You should also check with your local suppliers or contact the following trade associations: The National Wood Window and Door Association (708-299-5200) consists of window and door manufacturers. The National Sash and Door Jobbers Association (708-299-3400) consists of window and door distributors.
— *Mark Feirer, editor of* Fine Homebuilding.

Andersen Windows, Inc.
100 Fourth Ave., North Bayport, Minn. 55003-1096
(800) 426-4261
Low-e, argon-filled round-top and elliptical units. Wood or vinyl-clad. Extension jambs available. Laminated and curved trim available.

Atrium Door & Window Co.
P. O. Box 226957, Dallas, Texas 75222-6957
(800) 527-5249
Half-rounds, prefinished (primed line will soon be available), true divided lite or snap-in grills. High-performance glazing (HPG).

Caradco
P. O. Box 920, Rantoul, Ill. 61866
(217) 893-4444
Half-round, elliptical and quarter-round windows. Custom windows also available. Aluminum-clad or primed wood exteriors. Natural wood interiors. HPG.

Crestline
One Wausau Center, P. O. Box 8007, Wausau, Wis. 54402-8007
(800) 552-4111
Stock and custom. HPG. Wood and clad.

DashWood Industries, Ltd.
Box 10, Centralia, Ont., Canada N0M 1K0
(519) 228-6624
Aluminum-clad, vinyl-clad, encapsulated. Bare wood and primed.

DF Windows
Donat Flamand, Inc., 90, Industrielle St., Saint-Apollinaire, Que., Canada G0S 2E0
(418) 881-3974
Wood round-tops and half-rounds.

Hurd Millwork Co., Inc.
575 S. Whelen Ave., Medford, Wis. 54451
(800) 2BE-HURD
Standard sizes in half-round, quarter-round and ellipse. Custom sizes and shapes available (aluminum-clad or primed wood). HPG.

JJJ Specialty Co.
113 27th Ave., N. E., Minneapolis, Minn. 55418
(612) 788-9688 or (800) 445-6736
Wood and aluminum-clad windows. Elliptical and round-top windows. True divided lite or single lite with grill. Custom sizes only.

Kolbe & Kolbe Millwork Co., Inc.
1323 South 11th Ave., Wausau, Wis. 54401
(715) 842-5666
Round-top, half-round windows. Laminated jambs. Stock and custom. True divided lite and single lite with grill. Any wood species available in custom line.

Lincoln Wood Products, Inc.
P. O. Box 375, Merrill, Wis. 54452-0375
(715) 536-2461
All units are custom. Aluminum-clad (four colors) or unfinished. Round-top or quarter-round. HPG. Tinted or tempered glazing available.

Loewen Windows
1397 Barclay Blvd., Buffalo Grove, Ill. 60089
(800) 245-2295 or (708) 215-8200
Stock and custom. Wood and clad. HPG.

Marvin Windows
P. O. Box 100, Warroad, Minn. 56763
(800) 346-5128
Custom and standard round-tops. Aluminum-clad or unfinished. True divided lites or single lite with grill. Prefinished in standard colors.

New Morning Windows, Inc.
11921 Portland Ave. South, Burnsville, Minn. 55337
(612) 895-6175
Custom with any glazing. Wood and clad.

Norco Windows, Inc.
P. O. Box 140, 811 Factory St., Hawkins, Wis. 54530-0140
(800) 526-3532 or (800) 826-6793 (Wis.)
Half-round, custom and stock windows. Unfinished or primed wood exterior.

Peachtree Doors and Windows, Inc.
Box 5700, Norcross, Ga. 30091
(800) 477-6544
Custom. Aluminum or wood. HPG.

Pella Windows & Doors/Rolscreen Co.
102 Main St., Pella, Iowa 50219
(515) 628-1000 or (800) 524-3700
Stock round-tops. Aluminum-clad and finished or unfinished wood. Standard and custom colors.

Pozzi Window Co.
P. O. Box 5249, Bend, Ore. 97708
(800) 821-1016
Custom and stock half-rounds. Wood (single lite or true divided lites) or wood-clad. HPG.

Wenco
P. O. Box 259, W. Main St., Ringtown, Pa. 17967
(800) 255-7743
Stock and custom units. Aluminum-clad or unfinished wood. HPG and tempered glass available.

Zeluck, Inc.
5300 Kings Highway, Brooklyn, N. Y. 11234
(718) 251-8060, ext. 89
Custom windows only. Various woods, including mahogany, teak, walnut and cherry. HPG.

---

through our Williams & Hussey molder/planer in the past, but this casing was different. Making molding requires that the stock be fed straight into the cutterhead. With simple, curved pieces, the usual technique is to register the stock against curved guides. But the curves on the eyebrow casing reversed direction twice. Garry's low-tech solution was to free-hand the blank through the molder/planer by pivoting it on two wood blocks that he has screwed to an auxiliary melamine bed (bottom right drawing, p. 51). Garry positioned the blocks on the infeed side of the cutterhead, letting him use the blocks as he would the guide pins on a shaper table. The result wasn't furniture grade, but it was up to snuff as exterior architectural millwork. The casing was applied to the window, everything was sanded and primed, and the hardware was affixed to the operating sash. We were ready for installation.

Framing the dormer—At this point, Pascucci took over. Because the window would look out from an empty attic, the only headered opening needed was behind the operable hopper sash; full rafters would run behind the rest of the unit (for more on framing a full eyebrow dormer, see *FHB* #65, pp. 80-84). All the rafters above the window were painted black so that they wouldn't be visible from the street. Pascucci's framing technique let him support the window on seat cuts made in the top edge of the existing rafters (top drawing, facing page). Using a pine batten, Pascucci then drew in a fair curve corresponding to the intersection of the eyebrow's rafters with the main roof deck (photo facing page).

The dormer framing consisted of short 2x4 rafters that were screwed to the window's head jamb at one end and to the main roof at the other end. Sheathing this was a bit tricky. Because the

¾-in. plywood we used elsewhere wouldn't make the necessary bends, Pascucci used three layers of ¼-in. lauan instead. The lauan was applied in overlapping strips because full sheets could not easily conform to a compound curve. Fortunately, our roofer was able to blend the eyebrow dormer's cedar shingles into those of the main roof. The shingles were applied over an elastic-sheet membrane to keep any water that backed up under the shingles from leaking into the attic. At the junction of the two roofs, copper step flashing was hidden between courses in what was, in essence, a woven valley. This meant carefully choosing pliable shingles and, when that wasn't enough, boiling them. □

David Frane is a foreman with Thoughtforms Corp., a construction company in W. Acton, Mass. Photos by the author except where noted.

Installing Arch-Top Windows

How one builder supports the loads without a conventional header, then uses a trammel jig to cut siding and casing

by Douglas Goodale

Arch-top windows have caught the fancy of home buyers, so carpenters have to contend with the difficulties of framing and finishing these semicircles within rectilinear walls.

Round-top, circle-top or arch-top, whatever you call them, such windows have become a popular architectural feature. They add light and view and a touch of grandeur to a house. But these windows require carpenters to come up with some resourceful solutions when it comes to installation and trimming. Here, I'll describe techniques that work for me, including some I used on a recent project in rural Hunterdon County, New Jersey (photo above).

Solving the structural problems—A typical rectangular window has a structural header at the top of the rough opening, and the header is supported by trimmer studs that pick up any loads above. This header—I usually use a doubled 2x10 with a 2x4 laid flat on the bottom—sits below the double top plates, which means the top of the window is about 13 in. below the ceiling.

With arch-top windows, however, it's common to want the top of the window closer to the ceiling, which means there's no room for a conventional header. Usually the arch top will stop 5 in. or 6 in. below the ceiling line,

so that after installing the casing, you still have some clearance between the casing and the ceiling.

I use two methods to eliminate the conventional header. The first method is used if the window is in the first-floor wall of a two-story house. In this case, second-floor deck, wall and roof loads must be supported. I move the header up above the top plate (bottom drawing, facing page). In 2x4 framing, this would call for a double 2x10 with ½-in. plywood sandwiched between; in 2x6 framing, it would call for a triple 2x10 with ½-in plywood. In either case, I use the rim joist as the outside layer of the header. If the ends of floor joists bear above windows, I then install metal hangers to support the joists that meet at the header.

I use a different method if the window is in a wall supporting only ceiling and roof loads, where we often have less depth to work with. I usually have room for a smaller dimension header that's beefed up to carry the roof loads. For window openings 3 ft. to 4 ft. wide, I use a piece of ½-in. by 5½-in. steel sandwiched between 2x6s, or a piece of 4-in. by

4-in. steel angle (such as masons use for lintels) packed with two 2x4s. I bolt these headers together with ⅜-in. carriage bolts.

Framing and cathedral ceilings—On the house shown here (photo left), all the arch-top windows were in gable-end walls of rooms with cathedral ceilings. Cathedral ceilings present special problems because their construction usually requires a structural ridge beam. You cannot build a cathedral ceiling with conventional pairs of rafters opposing at a ridge board. Without ceiling joists or collar ties, the weight of the roof will push the outside walls apart. By making the ridge beam structural, however, the weight of the roof bears on the outside walls and on the ridge beam.

If there are no windows centered in the gable wall, the ridge beam is usually supported by a solid or built-up post, or by a header distributing loads to a pair of posts. When you place an arch-top window in the center of the gable, however, there is no room for the post and rarely is there room for a conventional header. Let's consider, for example, the 8-ft. dia. arch top I installed in the master bedroom. To support the ridge beam (a 5½-in. by 13-in. glulam) while keeping the arch-top window close to the peak, I turned the gable-end rafters into a truss.

First I made a pair of gable rafters out of 2x10s (the other rafters were 2x8s). I butted this first pair of gable rafters together in full plumb cuts. Then I fit a 2x12 crosspiece to the underside of the rafters so that the bottom edge of the crosspiece was even with the top of the rough opening for the arch top (top drawing, facing page).

Next I sheathed the inside face of this assembly with plywood, starting with a triangle whose sides followed the top edges of the rafters and the bottom edge of the crosspiece. I filled out the remaining faces of the rafters with ripped lengths of plywood.

Then I added another layer of 2x stock. But this time I ran a 2x12 chord all the way through—flush with the top edges of the rafters—and filled in with 2x10 stock above and below it. At the ridge, I notched the plumb cuts to create a pocket for the ridge beam. Next I added another layer of plywood, and finally, a third layer of 2x stock. I used construction adhesive between all the layers, and

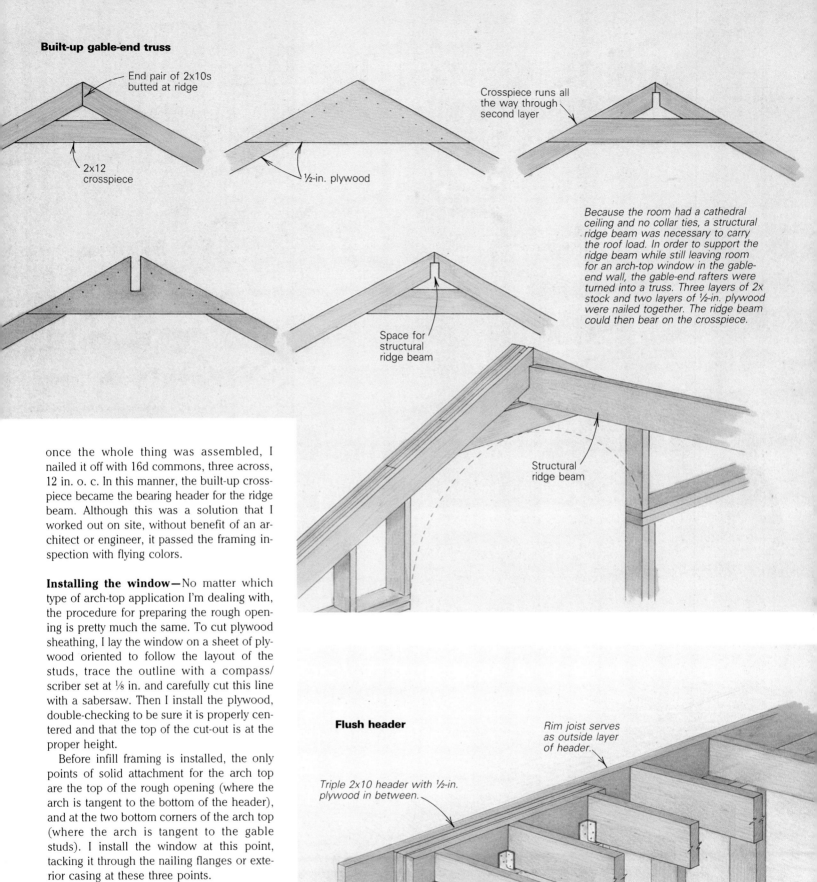

Built-up gable-end truss

End pair of 2x10s butted at ridge

2x12 crosspiece

½-in. plywood

Crosspiece runs all the way through second layer

Space for structural ridge beam

Because the room had a cathedral ceiling and no collar ties, a structural ridge beam was necessary to carry the roof load. In order to support the ridge beam while still leaving room for an arch-top window in the gable-end wall, the gable-end rafters were turned into a truss. Three layers of 2x stock and two layers of ½-in. plywood were nailed together. The ridge beam could then bear on the crosspiece.

Structural ridge beam

once the whole thing was assembled, I nailed it off with 16d commons, three across, 12 in. o. c. In this manner, the built-up crosspiece became the bearing header for the ridge beam. Although this was a solution that I worked out on site, without benefit of an architect or engineer, it passed the framing inspection with flying colors.

Installing the window—No matter which type of arch-top application I'm dealing with, the procedure for preparing the rough opening is pretty much the same. To cut plywood sheathing, I lay the window on a sheet of plywood oriented to follow the layout of the studs, trace the outline with a compass/scriber set at ⅛ in. and carefully cut this line with a sabersaw. Then I install the plywood, double-checking to be sure it is properly centered and that the top of the cut-out is at the proper height.

Before infill framing is installed, the only points of solid attachment for the arch top are the top of the rough opening (where the arch is tangent to the bottom of the header), and at the two bottom corners of the arch top (where the arch is tangent to the gable studs). I install the window at this point, tacking it through the nailing flanges or exterior casing at these three points.

Then I fabricate a curved ladder of infill framing to provide nailing for the sheathing outside, and the drywall and casing inside (drawing, next page). I use 2-in. wide plywood for the curved "rails" of the ladder, then nail 5/4 by 3-in. blocks radially along the plywood every 6 in. to 8 in., except where I turn the blocks on edge so they don't stick out past the plumb-cut ends of the plywood.

I fit the ladders on either side of the rough opening, shim them snugly against the outside surface of the arch-top jamb and secure

Flush header

Rim joist serves as outside layer of header.

Triple 2x10 header with ½-in. plywood in between.

Joist hangers

Header is moved above the top plates to make room for arch-top window.

Drawings: Michael Mandarano

To provide nailing for sheathing, drywall and trim around arch tops, Goodale installs curved ladders made of plywood sides with blocks nailed between them.

With the cedar siding tacked in place on the trammel jig, Goodale pivots the swing arm and neatly trims the boards for a perfect fit against the arch-top window.

Installation details

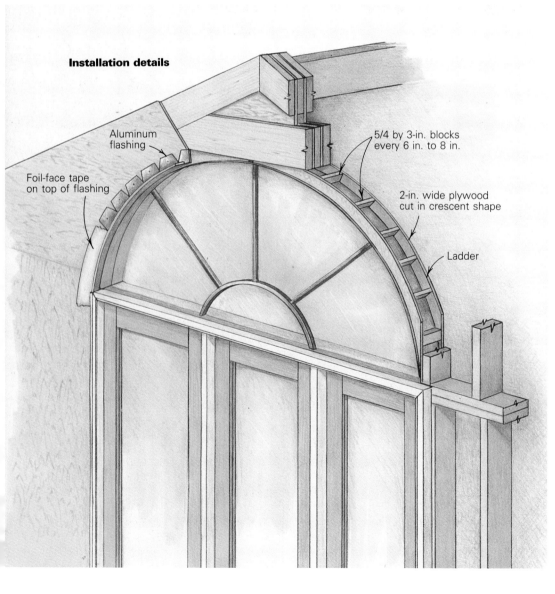

- Aluminum flashing
- Foil-face tape on top of flashing
- 5/4 by 3-in. blocks every 6 in. to 8 in.
- 2-in. wide plywood cut in crescent shape
- Ladder

them in place by screwing through the sheathing into the ladders. They provide continuous nailing inside and out, and also allow me to insulate around the window right up to the jamb. Rather than bang nails into the lightweight ladders, I run #6 1¼-in. drywall screws through the nailing flange of the window and into the plywood.

Curved flashing—Most vinyl-clad windows are designed so that the nailing flange acts as flashing. Metal-clad and all-wood windows are more of a problem. Some manufacturers offer vinyl flashing as an accessory that you can buy. Or you can have custom flashings made (out of copper, for instance), but that gets pretty expensive. I usually make my own flashing on site. Sometimes I'll even add this to a vinyl-clad arch top so that its flashing will match the windows and doors on the rest of the house.

I use a sheet-metal brake to bend 6-in. aluminum flashing into a Z-shaped profile, the dimensions of which vary depending on the window. In this case the bends measured 4⅜ in. up the wall, 1⅛ in. across the top of the window and ½ in. over the edge of the window. Then I make cuts in the 4-in. edge, right up to the bend. The spacing of these cuts also varies with the radius of the arch top—the smaller the arch, the closer the cuts must be.

Another series of cuts about 2½ in. apart in the ½-in. flange allows the metal to bend to the shape of the arch top. When fitting the flashing, a helper supports the uncut end while I make the cuts in the ½-in. flange and fit it to the trim. The flashing is attached by nailing through the 4-in. flange

Photos this page: David Schiff

with roofing nails, using one nail per tab. I've tried cutting the flange on a bench, but it makes the stock too flexible to handle without distorting. As an alternative to cutting the ½-in. flange, you can crimp it with a three-leaf crimper (used by sheet-metal workers and woodstove installers).

I know that the cuts in the flashing are a vulnerable area, so I use a good quality foil-faced tape to seal the joint where the wall and window meet. I tear off short lengths of tape, and starting at the bottom, overlap the pieces as I work my way up the window. Lastly, I run a generous bead of clear caulk before installing siding. So far, none of my installations has leaked.

Using the trammel jig—My trammel jig is a worktable with an adjustable arm that swings arcs of any radius from 6 in. to 60 in. I use it as a giant compass to lay out half circles on plywood, and with a router mounted on the arm, I use it to cut siding and molding for arch tops.

To build the trammel jig, I start with a full sheet of plywood and add smaller sheets to the two bottom corners and the top edge of the jig. The corner panels are 16 in. by 24 in. The top panel is 16 in. by 48 in. These extension panels are attached to the full sheet with 1x3 cleats, glued and screwed through the back.

The adjustable arm is a length of ½-in. plywood, 4 in. wide by 60 in. long. It has a centered 1-in. hole 2 in. from one end and a ⅝-in. wide slot cut down the length, stopping 2 in. short of the 1-in. hole and 2 in. short of the other end. The router is centered over the 1-in. hole and screwed to the plywood.

The arm pivots on a carriage bolt running through the base of the jig and through the slot in the swing arm. The radius of any arc is determined by the position of the cross-piece, which has a hole drilled in it for the carriage bolt, and which can be screwed to the arm at any point along the slot. It's important that the pivoting end of the arm be shimmed up the thickness of the stock being cut.

On this house, the T&G cedar siding runs square to the pitch of the roof. The trammel jig allowed us to cut all of the pieces for the window at the same time so they fit perfectly. As a guide for the top edges of the stock, I tacked two boards to the table at the roof pitch. To use the jig, I tacked the stock to the trammel jig, carefully keeping the nails out of the line of cut.

With the stock positioned in the jig, I mounted a pencil in the swing arm and marked the line of cut on the stock. I removed the pieces and cut each one individually with a sabersaw to minimize the amount of work the router would have to do and to avoid blowing out the edges of the stock. Then I replaced the pieces on the jig and made the final cut with two or three passes of the router fitted with a ½-in. carbide straight cutter bit (top right photo, facing page).

A common location for arch-top windows is in the gable end of a room with cathedral ceilings, but this creates the problem of supporting a structural ridge beam. Goodale solved the problem with a site-built wood truss. The arched casing in the photo above was made on a trammel jig.

I also made interior casings with the trammel jig, beginning by mounting a pencil in the swing arm and drawing the inside and outside edges of the casing directly on the worktable. Then I laid out short sections of stock, overlapping each piece. The number of pieces needed to complete the arch varies depending upon the width of the stock—wider boards will require fewer pieces to complete an arc, but will also produce long run-out in face grain and should be avoided unless the casings are to be painted.

I joined the pieces by slot-cutting the matching ends and gluing them together with a spline. Once the pieces were glued up, I mounted the resulting polygon in the trammel jig, and using it as a compass again, re-drew the inside and outside lines of the casing on the stock. Just as with the siding, I cut the stock close to the line with a sabersaw and then remounted it in the jig for final passes with the router.

To avoid putting unnecessary nail holes in the face of the casing, I held stock in the jig with drywall screws run through the back. Once the inside and outside edges were cut, I simply nailed it up (photo above). But you can switch the router bit to any desired molding profile and shape the edges or even the face of the casing. □

Douglas Goodale is a builder in Frenchtown, New Jersey. David Schiff is a writer and an amateur builder who assisted with the writing of this article.

Photo: Ross Cameron

Building a Kitchen-Counter Bay Window

A big, hinged window connects this kitchen to the deck and lets in the breeze

by Tony Simmonds

When I remodeled the kitchen in my house, I had to replace the window over the sink. Amid my collection of building materials, I had a large, double-glazed sash that had been liberated from a previous engagement as a bedroom window. But it was too big for the opening. Pack rats like me are stubborn, though, and it wasn't long before I devised a way of using the big window—not by enlarging the opening, but by creating a new one a foot outside the existing kitchen wall (photo right).

There's a deck outside our kitchen, and the roof overhangs the wall where the old window was. So it seemed natural and easy to build a bay window with its top sheltered by the existing roof and its base resting on the deck. The dimensions of the window (which, in the way of recycled treasures, began to dictate the terms on which all design proceeded) meant that the sill of the bay window would be 39 in. off the floor. I'm tall, and I could use a sink a little higher than that. But the world is full of the vertically challenged, and those of us who are not must learn to compromise or be prepared to wash all the dishes.

So I decided that the counter and the windowsill would be one and the same: The counter surface simply would flow through the wall and become the sill of the bay. There was an advantage to be gained from this plan. Space is constrained in that part of the kitchen, and there isn't really room for a standard 25-in. deep counter. Rather than crowd a small sink into a counter too shallow for it, I could borrow space from the wall cavity under the counter, and I could repay it by adding insulation and a layer of sheathing to the outside of the existing wall because this new space would all be inside the base of the bay window. In fact, there would still be room inside this base for shallow storage shelves, accessible from the deck outside (top drawing, p. 60). Beyond these practical advantages, I was excited by the idea that

A light-filled bump-out. Tucked in the shelter of the eave, this bay window extends the kitchen counter past the sink and provides shelves for plants. The 4-ft. by 5-ft. window swings open, awning style. On the exterior side, the sliding doors below the window conceal the barbecue gear.

this window might be made to open. A single sash 4 ft. high by 5 ft. wide would have the kind of wall-abolishing effect that French doors do, making the kitchen counter an inside and outside space (left photo, facing page). Such a window could only open upward. Hanging it and fastening it promised to be a challenge, but I was sure it was a challenge that could be met.

First, the sink base—The highest priority in every kitchen renovation I've done is to get the sink functioning again, and my own was no exception. I stripped the wall to the sheathing, inside and outside, and put in the bare bones of the sink cabinet—just enough to get the counter in place.

Inside, over the kitchen cabinet, the plywood substrate is fastened in the usual way with screws through the cabinet stretchers. Outside the wall in the bay, I screwed the plywood to a 2x4 frame supported at either end by short walls. The walls are anchored to the deck with 12d nails and to the sheathing with screws driven from the inside. I biscuit-joined and glued the joint between the two pieces of plywood to ensure alignment and stability.

The counter is finished with square tiles of mossy green Chinese slate. But before the slate could go down, I had to install a wood border to contain the slate along the edge of the counter and the perimeter of the bay window. I milled the edging from alder, an abundant and underrated hardwood with some of the visual characteristics of cherry, at about half the price. I screwed the edging to the plywood from underneath. Then I sealed the edging to protect it from mortar and grout stains with three coats of Behr Tung Oil Finish (Behr Process Corp., P. O. Box 1287, Santa Ana, Calif. 92702; 714-545-7101).

Framing the bay—With the counter finished and the kitchen sink back in operation, I could concentrate on the window. The first step was to get the sill in place. I milled it from 6/4 Douglas fir, which meant that in order to maintain a solid-looking 1-in. thickness at the outer edge, I could bevel the top surface only at about 5°. Typically, I put a 10° bevel on a windowsill, but the shallower bevel is fine here because the entire window is sheltered by the overhang of the roof.

To support the sill, I ran a 2x ledger around all three sides of the counter-platform framing at a height that would result in a ½-in. reveal between the top surface of the sill and the alder edging, which acts as a window stool at this part of the bay. I biscuit-joined and glued the sill at the mitered corners with Titebond Type II glue, a fast-setting, one-part yellow glue that passes Type II

A counterweight holds the window up

To help hoist the window, the author installed a counterweight (photo bottom right) that connects to the sash by way of steel cables. The cables are threaded through sheave blocks secured to the wall and soffit above the window and to the trellis over the deck. When the window is closed, the cable ends are detached and hooked together by screw-open chain links.

Screw-open chain links

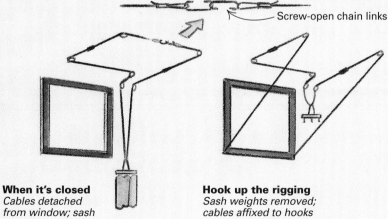

When it's closed
Cables detached from window; sash weights in place.

Hook up the rigging
Sash weights removed; cables affixed to hooks recessed in window-sash corners.

Lift the window
Sash weights in place; window ready to be opened.

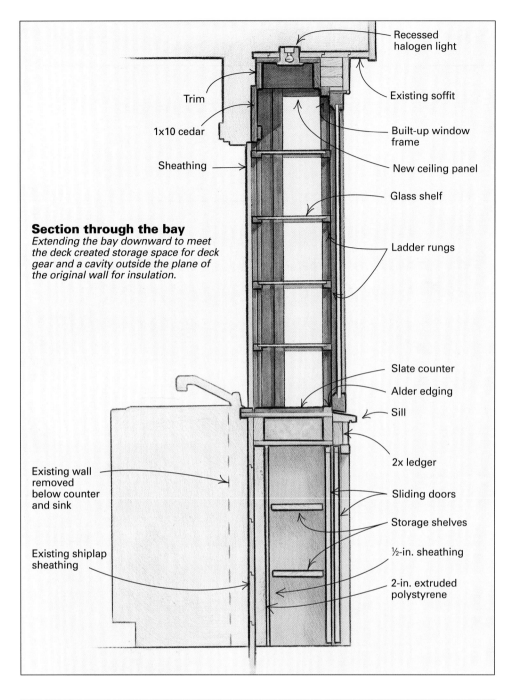

Section through the bay

Extending the bay downward to meet the deck created storage space for deck gear and a cavity outside the plane of the original wall for insulation.

Recessed halogen light

Existing soffit

Trim

1x10 cedar

Built-up window frame

Sheathing

New ceiling panel

Glass shelf

Ladder rungs

Slate counter

Alder edging

Sill

Existing wall removed below counter and sink

2x ledger

Sliding doors

Storage shelves

Existing shiplap sheathing

½-in. sheathing

2-in. extruded polystyrene

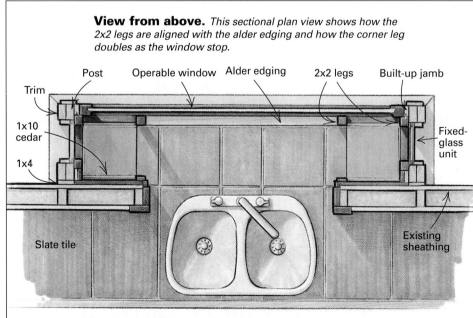

View from above. *This sectional plan view shows how the 2x2 legs are aligned with the alder edging and how the corner leg doubles as the window stop.*

Post

Operable window

Alder edging

2x2 legs

Built-up jamb

Trim

1x10 cedar

1x4

Fixed-glass unit

Slate tile

Existing sheathing

testing for water resistance. I fastened the sill to the ledger with 4-in. #14 screws, 12 in. o. c. Then I plugged all the holes, even though some of them later would be covered by the corner posts.

After laying out the corners of the window on paper, I decided that there was insufficient space for both rough framing and a separate jamb. On the other hand, the proportions I thought were needed for the finished post would have required a solid post to be glued up and then milled in a complex and detailed way. The solution I arrived at is a hybrid. It has a structural core consisting of a single, kiln-dried spruce 2x4, ripped down to 3 in. wide; the core is wrapped with trim and a two-piece built-up jamb (drawing bottom left). This design allowed me to make the corner posts out of smaller dimension pieces, most needing to have only one good face and all needing only simple end cuts.

I transferred my layout from drafting paper to the sill, and then I used a plumb bob to locate the corners on the soffit above. I needed to build down 5½ in. from the soffit to the top of the sash. Allowing ¾ in. for the jamb, plus some clearance, I figured three layers of KD 2x4 would get me about where I needed to be (drawing top left). Because I planned to fasten the two-part jamb directly to this head framing, however, I needed to level the bottom layer of 2x4 carefully. Having already taken care to level the sill, I simply cut the 2x4 corner posts to length and then shimmed down the bottom 2x4 to meet them. I used exterior-grade screws for all fastening. Screws are less likely than nails to pull things out of whack, and by predrilling, even toe-screwed connections can be made accurately and securely without fear of splitting the wood.

Because I wanted a 3-in. wide vertical trim to show where the bay meets the wall, I had to build another post there, too. I made up an L with a 1x4 flat to the wall and a 2x4 perpendicular to it—both of them ripped to 3 in. wide—and fastened it to the wall through the 1x leg (drawing bottom left). Next I doubled up each 2x4 post with a second piece of KD spruce, selected for appearance on its outer face and screwed from the back so that no fasteners showed.

Next came the 2x2 component of the built-up jamb. All along, I had seen the 2x2s as getting drilled for shelf supports so that we could put small shelves in the recesses of the bay. But as the window took shape, I was attracted by the idea of ladders to support the shelves (photos facing page). I experimented with different layouts for them, and by happy coincidence I found that an equal division of the vertical space into five brought the three bars closest to eye level into nearly perfect alignment with the open shelves on the inside of the kitchen wall. From the outside, the effect is of a mullion and horizontal muntin bars breaking up the expanse and height of the window.

At the ends of the bay, the 2x2 legs act as stops for narrow fixed-glass windows (drawing bottom left). And at the corners and along the top, the 2x2s serve as stops for the operable window. My next task was to graft onto the 2x2s the ¾-in. by 2-in. portions that complete the built-up jamb. After checking that the framed opening was square and correctly dimensioned, I used a biscuit joiner to cut slots in the edge of the prepared stock and in the outside face of the 2x2s. Before glue up, I routed the hinge gains in the header piece. After glue up, I reinforced the head jamb by running a 2-in. #12 screw through each hinge gain into the header framing.

Rig the sash with a counterweight—Hanging the sash brought to a head the problem I had anticipated without quite knowing how I was going to solve it. This sash weighed close to 75 lb. How were we going to raise it to its open position? And once having raised it, how would we make sure it stayed there instead of crashing down and killing somebody standing underneath it?

I had entertained suggestions from all comers while being fairly sure that some form of counterweight was the only answer. In anticipation of this situation, I had positioned one of the trellis members on the deck outside directly over where the bottom rail of the open sash would be. I thought I could recess a screw eye into each bottom corner of the sash so that the cable connecting it to the counterweight could be detached and retracted out of the way when the window was closed.

There was still a problem, though: How would I draw down the cable ends with a 30-lb. counterweight attached to their other end when I was ready to hook them up and open the window? And then how would I avoid having the same 30-lb. counterweight crash to the deck when I detached the cable ends to close the window? Our friend Dick Fahlmann, who has built and rigged movie sets for 20 years, suggested the principle of a changeable counterweight.

There are a number of ways of making an adjustable counterweight. I chose a simple, crude system that employs sash weights from old double-hung windows attached with S-hooks to an old cast-iron ring courtesy of the Vancouver Water Works (photo bottom right, p. 59). Removing the four weights around the perimeter is enough to make it easy to pull down the cable ends, which are fitted with screw-open chain links that I found in a pet store (they have a 440-lb. breaking strength) (drawings p. 59). Once these links are hooked onto the 2½-in. screw eyes in the sash corners, you simply replace the weights. The window rises easily.

In fact, there is enough friction in the system that nothing happens too suddenly. A gentle pull on the counterweight is required to raise the window all the way, and once raised, there's plenty of inertia to keep it there safely. The procedure

Rabbeted rungs support glass shelves. The 2x2 ladder legs support the rungs on two sides. On the exterior of the old wall, the rungs are sandwiched between courses of clear, red-cedar siding.

takes half a minute or so to accomplish (photo top right, p. 59).

I used ⅛-in. galvanized aircraft cable and galvanized fittings throughout. The sheave blocks are inexpensive fish-boat tackle, which Dick located for me. As well as costing far less than stainless steel, galvanized has another advantage: It lets you know when it's wearing out, unlike stainless, which looks like new right up to the moment it breaks. Whichever you choose for any rigging job, it's important to use the same or compatible metals throughout to avoid galvanic corrosion. The sleeves that guide the cable through the fascia are lead screw-anchors with their bottoms drilled through. Lead and galvanized steel have no corrosive effect on one another.

Finishing up, inside and out—After installing the fixed glass in the ends of the bay and trimming out the corner posts, I finished the ends of the cupboard underneath the bay in frame-and-panel style, and I built sliding doors for the front of it. These doors are 1x6 resawn cedar stained to match the siding on the deck's guard walls. I made the boards ½ in. thick and milled a shiplap edge on them. Then I fastened them with brads and a bead of construction adhesive to some ½-in. plywood. The stiles were left full thickness and rabbeted to cover the edge of the plywood. The doors are hung with inexpensive bypass-door hardware.

To light the bay, I used two low-voltage halogen lights equipped with 50-watt bulbs. I first installed only one of these, centered over the bay, but the 40° spread of the floodlight left the shelves in the recesses poorly lit. Because I had experimented by installing the light in the existing soffit before I put in the finished ceiling, all it cost was money and another trip to the supplier to change my mind.

Rather than cut and nail individual pieces in the cramped space of the bay ceiling, I made a panel of 1x4 cedar, biscuit-joined and glued edge to edge, with a V-joint to match the look of the tongue-and-groove soffit outside. This echoing of the rough exterior finish with a polished wood finish of similar proportion was what I tried to achieve with the wide cedar boards I put on the short sections of wall between the bay ends and the original window opening. Although these walls are inside, they are seen only from outside. By lining up shingle courses with ladder rungs, the connection—and contrast—between outside and inside is highlighted and intensified. □

Tony Simmonds operates Domus, a design/build firm in Vancouver, B. C., Canada. Photos by Charles Miller except where noted.

Framing a Walk-Out Bow Window

A full-scale drawing reduces the chance of error

by Carl Hagstrom

When the subject of windows came up during the design stage of a house I built recently, the homeowner handed me a page taken from a magazine that showed a walk-out bow window. "That," she said, "is the window I want in my living room."

And why not? Her house was sited on a hillside overlooking a big chunk of farmland, and with a window like this, she'd get a real eyeful of the landscape.

While all of the thick talk of geometry can make a bow window seem daunting, it's really not that difficult to frame and install. The key is to make a full-scale drawing. I used the manufacturer's specs to frame the rough opening, and when the window was actually on site, I set it up on the floor, snapped a few reference lines and made a full-scale drawing of the window and the framing layout. The window fit beautifully (photo right), and I didn't have to fool with a lot of geometry.

Is it a bow or a bay?—The terms bow window and bay window are often mistakenly used interchangeably. But they're two different styles of projecting windows. A bay window has three sides, with the center portion parallel to the wall plane. The other two sides can return to the wall at various angles (90°, 45°, 30° being common). A bow window, on the other hand, can have any number of sides, with each side tangent to an arc of a circle.

Both window styles commonly employ a seat at the bottom of the window to enclose the lower area of projection, but to become a walk-out unit, the floor must extend out to meet the profile of the window projection.

Homework is required—From experience, I knew that a walk-out bow window would present some framing challenges—like where and how to cut the floor joists, how to transfer loads to floor joists and what to do above the window. So after I ordered the window, a 6-ft. high by 10-ft. wide Andersen unit

(Andersen Corp., 100 4th Ave., North Bayport, Minn. 55003-1096; 612-439-5150), I read Andersen's literature and firmed up a framing plan.

I decided to frame the rough opening, then leave the rest until the window was on site. Andersen's literature said to make the rough opening 10 ft. ³⁄₁₆ in. by 6 ft. 1⅞ in. The rough opening is where the nailing flanges of the end windows hit the wall plane. I treated this opening as I would any rough opening: with a header resting on trimmer studs. In this case I spiked three 2x10s together to span the opening and to support the roof load.

To create the walk-out projection, I cantilevered the wood I-beam floor joists over the founda-

tion wall. According to Andersen, the window projects 14 in. from the exterior wall plane. I cantilevered the joists about 2 ft. beyond the wall plane and cut them to exact size and angle later.

I was installing this bow window in a one-story home that had 20-in. deep eaves, so I could tuck the window projection under the roof overhang. Usually, bow and bay windows have their own roofs, like tiny additions to a house. The house's style governs a bow window's roof design, so if you're considering either a bow or bay window, and there's no overhang to tuck it under, the roof over the window deserves some thought.

Although bow windows are commonly shipped fully assembled, I had the window shipped in pieces because I didn't want to lug around a 400-lb. window unit. This bow window consists of five standard windows. To achieve the bow, Andersen provides wedge-shaped spacers that are screwed between the windows. I assembled the bow window temporarily inside the house, where I could snap some lines and figure out the framing layout.

Making the drawing—With the window unit temporarily assembled in the living room, I struck two reference lines on the subfloor: an interior-wall line 6 in. from the nailing flanges on the end windows and a perpendicular centerline (top drawing, facing page). Then I measured the rough opening and checked it against the reference lines to be sure the window unit would fit. It did, which meant the window was in place—theoretically at least.

The next step was to redraw these two reference lines on a sheet of drywall (middle drawing, facing page). Then with a framing square and a T-square, I drew the footprint of the bow window on the drywall by transferring measurements from the windows to the drywall, using the reference lines as guides.

I also traced the wall framing. It was easy enough to measure from the centerline and draw the 2x6 jack studs at both sides of the rough opening. But the window framing

Windows from floor to ceiling. This 6-ft. high by 10-ft. wide Andersen walk-out bow window comprises five standard casement windows joined with wedge-shaped spacers. The unit projects 14 in. from the wall plane and is supported by cantilevered wood I-beam floor joists.

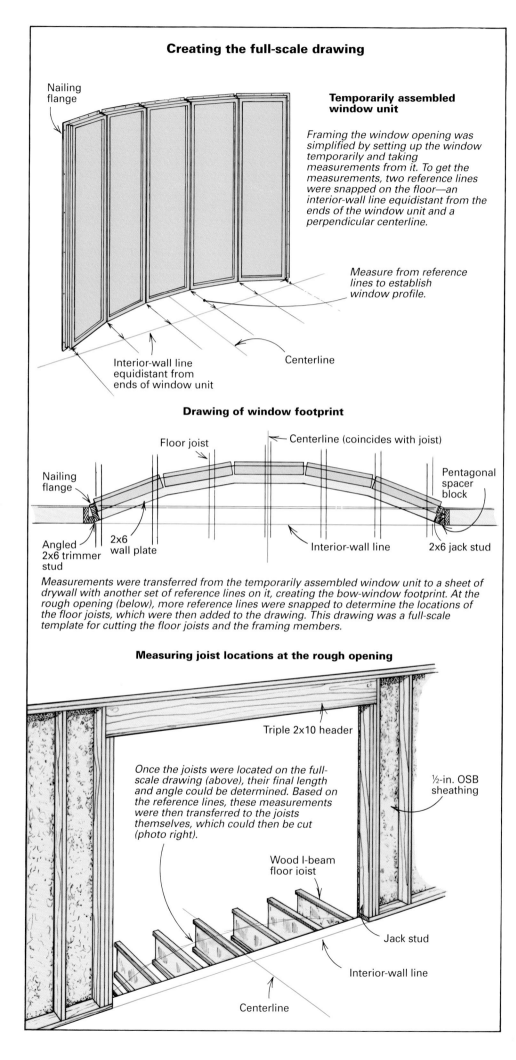

Creating the full-scale drawing

Nailing
flange

**Temporarily assembled
window unit**

*Framing the window opening was
simplified by setting up the window
temporarily and taking
measurements from it. To get the
measurements, two reference lines
were snapped on the floor—an
interior-wall line equidistant from the
ends of the window unit and a
perpendicular centerline.*

*Measure from reference
lines to establish
window profile.*

Interior-wall line
equidistant from
ends of window unit

Centerline

Drawing of window footprint

Floor joist

Centerline (coincides with joist)

Nailing
flange

Pentagonal
spacer
block

Angled
2x6 trimmer
stud

2x6
wall plate

Interior-wall line

2x6 jack stud

*Measurements were transferred from the temporarily assembled window unit to a sheet of
drywall with another set of reference lines on it, creating the bow-window footprint. At the
rough opening (below), more reference lines were snapped to determine the locations of
the floor joists, which were then added to the drawing. This drawing was a full-scale
template for cutting the floor joists and the framing members.*

Measuring joist locations at the rough opening

Triple 2x10 header

*Once the joists were located on the full-
scale drawing (above), their final length
and angle could be determined. Based on
the reference lines, these measurements
were then transferred to the joists
themselves, which could then be cut
(photo right).*

½-in. OSB
sheathing

Wood I-beam
floor joist

Jack stud

Interior-wall line

Centerline

also turned out from the wall framing 20°—the
window's starting angle, which I got from Andersen's literature—so spacer blocks and trimmer
studs were necessary.

On the drawing, the trimmer studs follow the
angle of the end windows—20°—and touch both
the nailing flanges and the front edges of the jack
studs. Then to complete the interior wall plane, I
drew pentagonal spacer blocks that fill the 20°
openings between the trimmer and jack studs.

The last step was to add the locations of
the cantilevered floor joists to this drawing. I
snapped another pair of reference lines at the
rough opening itself (bottom drawing, left). Then
I transferred the location of the floor joists to the
full-scale drawing, once again using the framing
square and the T-square.

Cutting the floor joists—Referring to the drawing, I laid out and cut the cantilevered floor joists.
I measured from the reference line to the ends of
the joists, subtracted the thickness of the sheathing and drew the angle of the cut on the top of
the joist. Then I squared a plumb line down the
face of the joists and cut each joist.

Unlike framing lumber, the engineered joists
that I used have an I-beam profile that makes
them difficult to cut with a circular saw. So I
padded the web to match the flanges with a temporary 1x3 block screwed next to the cutline.
This block provided a flush surface for the saw
table to ride on (photo below).

At both ends of the bow, the joists project only
a few inches; I used a reciprocating saw to cut
these short joists. Cutting with a reciprocating
saw requires a steady hand, but it was the best
tool to use in such a confined space.

I also used the drawing to cut 2x blocking that
spans the bottom flanges of the joists. This blocking does two jobs. First, because the window
junctions do not occur directly over the floor
joists, the blocking extends the profile of the bow
window down to the bottom of the joists. It also
provides backing for the sheathing.

Next, I marked and cut the plywood subfloor
and glued and nailed it to the floor joists. I also

Cutting the cantilever. **Floor joists (wood I-
beams on 19¼-in. centers) at the rough opening were left long and then cut to the bow
shape. The exact lengths of the joists and angles of the cuts were taken from the drawing.**

Taking shape. After cutting the joists, 1x4 blocking was nailed along the bottom of the joists to continue the window profile. Then the plywood subfloor was glued and nailed to the joists, and the 2x6 wall plates were measured right from the shape of the subfloor and nailed off.

Spanning from joist to joist. The window junctions, which are the weakest points of the unit, don't occur over the floor joists, so doubled 2x8 headers were installed to transfer window loads to the floor joists. These headers are positioned so that they sit within the 2x6 wall plate.

Installing the sill. The walk-out bow window rests on a 2x6 sill; each piece of the sill is identical to the wall plate below it. A level held against the plate indicates when the sill is plumb.

cut and installed a piece of ½-in. OSB on the underside of the cantilevered floor joists.

Because the outside edge of the walk-out represents the exterior surface of the wall framing, I was able to lay out the 2x6 wall plates by taking measurements from the outside edge of the subfloor (top photo, left).

Hidden headers—Because the bow-window junctions don't land on the cantilevered floor joists, I installed doubled 2x8 headers below the window unit (middle photo, left). The headers transfer the window load to the floor joists and support the bow window at its weakest points: the junctions.

By adjusting the position of the headers within the 5½-in. thickness of the wall plane, I was able to maneuver each header so that it would bear on a pair of joists without projecting beyond the bowed wall plane. After the headers were toenailed in place, the windowsill plates followed, mirroring the wall plates below (bottom photo, left).

With the bowed wall plane as my reference, I used a level and a straightedge to transfer the layout of the wall framing to the roof-truss overhang above. Then I framed a 14-in. tall 2x6 wall that matched the profile of the bowed wall framing, lifted this 2x6 wall into place and nailed it off (top photo, facing page). Because the roof load is carried by the header in the rough opening, this wall is nonstructural and could be framed with studs. To complete the opening, I cut the pentagonal spacers on the table saw and nailed them in place, followed by trimmer studs that provide nailing on the sides of the window unit.

Installing the windows—After sheathing the upper and lower portions of the bowed wall, I installed the five windows. My coworker and I broke down the assembled unit and numbered each window and its accompanying wedge-shaped spacer so that all the screw holes would line up when we reassembled the window in place. By using the same screw holes, the window went back together the same way that I drew it on the drywall.

From that point on, installing the bow window was straightforward. The first unit was tacked in place, the wedge-shaped spacer screwed to the window frame (bottom photos, facing page), and the second unit was then screwed to the first unit. Following this sequence, all the units were fastened together but only tacked in place.

Then we pried and shimmed the bow window into position. When it was level, plumb and centered in the opening, we nailed off the window through the exterior flanges, spacing our 2-in. galvanized roofing nails 8 in. o. c.

The window unit came with vinyl weather caps that I snapped in place to provide a weather seal at each window junction. The caps got their first test right away because it started pouring as I was putting the last cap in place. ☐

Carl Hagstrom manages Hagstrom Contracting, a design/build company in Montrose, Pa. and is a frequent contributor to Fine Homebuilding. *Photos by Rich Ziegner.*

Framing over the window. Because of the house's deep roof overhang, a wall, not a roof, was built above the window. The wall consists of five separate 2x6 frames built to match the window profile. The frames were fastened together, and the assembled wall framing was nailed to the roof trusses and to the header spanning the rough opening. Later, the header will be padded down to the bottom of the 2x6 wall frame.

Fastening a spacer. A wedge-shaped pine spacer screwed to the window frame angles the windows to give the unit its bowed shape.

Nice view, eh? Against a backdrop of Pennsylvania farmland, the author checks a spacer for its number: When the bow window was first assembled, each wedge-shaped spacer was numbered so that everything would go back together according to the full-scale drawing.

Installing Glass Block

Privacy and light in the same package

by Michael Byrne

As a tilesetter, many of the jobs I've worked on over the years have included walls of glass block that were installed by others. Unlike bricks, masonry block and stone, which have rough textures and seem to suck up mortar and stay put, glass block slips and slides during installation, and refuses to behave unless treated correctly. Fed up with hours of making my tile fit around the lousy glass-block work I found, I began to learn that trade myself about eight years ago.

My first attempts to build a glass-block panel (a panel is a "window" of block surrounded by wall) seemed awkward and confounding, and before long, I began to sympathize with those whose work I had criticized. But I've done a number of glass-block jobs since and am finally confident in my skills. What follows are some tips that will help you to avoid the painful learning process I went through.

Characteristics of glass blocks—Glass blocks were originally made by hand (sidebar, facing page), but today's blocks are machine-made by pressing two molded halves of semi-molten glass together. The edges of the inner and outer faces create a flange that allows the block to key into the mortar joints, and allows space for metal reinforcing rods or wire. The raised bump formed at the point where the two halves are fused together is an additional mortar key. To reduce the slipperiness of glass when mortar is buttered on, the hidden edges of blocks are spray-coated with a plastic-like polyvinyl butyral coating while they are still hot from fusing.

The most common sizes of hollow glass block are 6x6, 8x8 and 12x12. Other sizes and shapes are available, including 4x8 and 6x8 rectangular blocks, and hexagonal corner blocks. Blocks manufactured in the U. S. are modular; that is, the nominal size includes an allowance for a ¼-in. mortar joint. The actual dimension of a modular block is ¼ in. less than the nominal size (a 6-in. block, for example, is actually 5¾ in. square). Though you can find a few glass artisans who will provide custom-made solid glass block, there's only one company in the U. S. that manufactures glass block—Pittsburgh-Corning Corporation (800 Presque Isle Dr., Pittsburgh, Pa 15239). Imported glass blocks are available (see the list on p. 69), but direct technical information and assistance may be limited.

In addition to the different sizes, Pittsburgh-Corning makes hollow block in two thicknesses. The standard block is 3⅞ in. thick and has an insulating value of R-1.96. It can be used for both interior and exterior applications. An 8x8 block weighs 6 lb. For light commercial and residential jobs, Pittsburgh-Corning makes a "Thinline" block that's 3⅛ in. thick, with an insulating value of R-1.75. Because Thinlines aren't as strong as the thicker blocks, their use is limited to exterior panels of 85 sq. ft. and interior panels of 150 sq. ft. But they are about 20% lighter than standard blocks, so they're ideal for situations where weight would be a problem. An 8x8 Thinline weighs 5 lb. Another kind of glass block is solid, instead of hollow. Called VISTA-BRIK, it's a solid chunk of glass that's suitable for translucent pavers between two levels of a structure, or as a nearly vandalproof panel. An 8x8 VISTABRIK weighs 15 lb., and can withstand a 30.06 rifle shot from 25 ft.

Standard glass blocks have a compressive strength of 400 psi to 600 psi, and solid glass VISTABRIK blocks are rated at 80,000 psi. Glass blocks, however, should not be used in load-bearing situations, regardless of the application or panel size. Provision must be made to channel loads around the block panels with properly sized lintels or headers.

Most of Pittsburgh-Corning's line can be specially ordered with inserts of translucent fiberglass that improve the R-value by about 5%. You can also get solar reflective block that cuts down on solar heat gain. In combination with these characteristics, various patterns can be pressed into the inside surfaces during manufacture, while the two halves are still semi-molten. These patterns change the way light passes through the block, and also change the amount of privacy blocks offer. Some blocks are clear, while others admit light but not the view. Putting the pattern on the inside of the block leaves the outside of the block smooth, making the blocks easier to clean once they're in place.

Mortar and admixes—Glass blocks are stacked one upon the other, and mortar joints are usually aligned both horizontally and vertically. You can stagger the joints if you wish, but this makes the blocks more expensive to install. Laying up glass block is like laying up concrete block or brick, except for one crucial difference: the consistency of the mortar.

Masons work with a rather moist mortar mix to compensate for the absorbent nature of brick and concrete block. When either one is laid up, moisture is wicked from the fresh mortar, which then stiffens and supports the weight of the masonry. Glass blocks, however, aren't absorbent, so standard mortar won't stiffen as quickly. Instead, it will ooze out of bed joints and head joints after a few blocks are in place—a sight greeted with considerable dismay by novices.

Mix your mortar on the dry side. Pittsburgh-Corning recommends a mix of 1 part portland cement, 4 parts clean sharp sand and ½ part lime (measure by volume). Add just enough water to change the mix from crumbly to spreadable—roughly five gallons of water to a 94-lb. sack of cement. The mix should just barely wet the edges of the blocks. (I prefer a mortar mix of 3 parts sand and 1 part cement, no lime, with a liquid additive in place of water. I'll discuss additives more below.)

You won't have a lot of working time with the drier mortar, so in moderate temperatures mix only what you'll need in about a half-hour; mix even smaller batches in very dry or hot weather. Don't add more water to mortar that begins to set up because this prevents it from curing properly. It also increases the likelihood of mortar cracking when it sets up.

You can use a pre-packaged mortar mix to which you add only water, though I prefer to mix my own. These mixes usually set up faster than a standard mortar, which can sometimes be an advantage. Pre-packaged mortar mixes have a disadvantage, too. They often contain a finer grade of sand than you get from a building-supply yard, so the mortar will shrink more as it sets. On small panels this won't present a problem, but on larger jobs you should use some sort of liquid mortar additive to control shrinkage.

The additive should also provide some waterproofing. As one side effect of waterproofing, the mortar gets stickier and so gets a better grip on the slippery blocks. An additive also increases the flexibility of the cured mortar joint, reducing cracking. The additive I use (which does all this) is Laticrete 8510 (Laticrete International Inc., #1 Laticrete Park North, Bethany, Conn. 06525). If you're making mortar from scratch in an area where special additives are hard to come by, you can get a sticky mix by using waterproof portland cement instead of standard portland.

A third kind of additive, called a mortar fortifier or lattice, can increase the compressive strength of the mortar. It will also make the mortar stickier and more flexible, and increase its water resistance after cure.

Panel design—The overall size of a panel and whether it's to be built on the interior or exterior of your house will determine the thickness of

In this shower surround, glass block was used as an interior partition and as a divider between the shower and the adjacent sunroom. The pattern inside each block obscures views yet admits light.

The history of plate glass and glass block is linked with the history of modern architecture. Early in this century, the classic precepts that underpinned much of architectural thinking were being challenged, and a fresh new vision was emerging in all the arts. This vision was due, in part, to the technical advances in construction materials and techniques that allowed architects more flexibility with structural design. Cast-iron columns could be spaced widely apart, and because of advances in glass technology, ever wider and longer sheets of plate glass began to fill the spaces between.

Glass block was another product whose invention depended largely on manufacturing advances. The "glass brick" walls in early modern architecture were actually single thicknesses of pressed glass squares set in large reinforced-concrete panels. These panels were either formed on site or precast before being installed as infill between structural elements. Auguste Perret's Notre Dame de le Raincy (1922) and Le Corbusier's Immeuble Porte Molitor (1933) provide examples of this technique.

In 1902, the Corning-Steuben Company invented the hollow, modular glass units that became known as glass block. The blocks were made by pressing molten glass into identical molds. The square "dishes" thus formed were sealed together at high temperatures, with a partial vacuum forming in the space between dishes as hot air cooled and contracted. Earlier attempts to form glass blocks using traditional glass-blowing techniques had proven unsatisfactory, because moist air from the lungs of the glass-blowers would condense on the inside of the blocks, clouding them. Owens-Illinois made glass block at one time, but with architectural preference for materials nearly as changeable as Paris fashions, they dropped out of the market years ago.

The directions for installing glass block offered to builders in 1902 were amazingly similar to the specifications offered today by Pittsburgh-Corning (formed by the merger of the Corning Glass Works and Pittsburgh Plate Glass Company), which is the sole U. S. producer of glass block. Back in 1902, Corning-Steuben said "The [glass] bricks are laid in a similar way to ordinary bricks, with a mortar consisting of one part portland cement, two or three parts fine sharp sand and one-fifth lime mixed not too thin with water." Contemporary Pittsburgh-Corning instructions go on to suggest the addition of a waterproofing admixture if waterproof portland cement mortar is not used.

The revival of interest in daylighting is one reason for the current wave of interest in glass block. Another factor, ironically, is a revived taste for historic allusion in architecture. European and Japanese manufacturers are now scrambling to catch a share of expanding sales in glass block.

Arthur Korn, in his 1926 book *Glass in Modern Architecture,* spoke for all the pioneers of modern architecture when he identified glass and glass block as altogether exceptional materials, "at once reality and illusion, substance and shadow."

—*Ronald W. Haase*

block to use and the method of installation. Though the design of glass-block walls can get complicated on large commercial installations, most residential jobs are pretty straightforward. For design purposes, these fall into two categories: small panels (under about 25 sq. ft. in area) and large panels (up to 144 sq. ft. in area).

Small panels, like a slender sidelite alongside an entry door or a glass-block divider in a shower room, add a touch of elegance and style to a house. From a technical standpoint, a small interior panel, like a block window between two rooms, is the easiest to build. Jobs this small can be mortared in tight at jambs and headers. For interior panels, no reinforcing is necessary, and you needn't waterproof the surrounding framework. But make sure that even a small job is adequately supported. Glass blocks may look light and airy, but a small panel of standard 12x12 blocks can weigh as much as 400 lb., not including the mortar. Put in extra cripples be-

neath the sill, or build up a heavyweight sill with an extra 2x to keep deflection to a minimum.

For a small exterior panel (and any other panel frequently exposed to water), the sills should be waterproofed before the block is installed. Pittsburgh-Corning doesn't call for waterproofing the jambs, but I often do it anyway. Tar paper embedded in asphalt emulsion is a suitable waterproofer for many situations. But I prefer to use a waterproof membrane (like Nobleseal, made by The Noble Co., 614 Monroe St., Grand Haven, Mich. 49417). Membranes are fairly easy to install, too; just wrap a sheet 2 in. or 3 in. over the framing and staple the edges in place; lap the seams and seal them with Noble adhesive. Even though Pittsburgh-Corning requires no reinforcing for a small exterior panel, I feel more comfortable about the job if it is reinforced. I'll talk more about reinforcing below.

For large interior panels, like a full-height partition between two rooms, the design gets a bit

Any glass-block panel exposed to water, such as this divider between a tub and a shower stall, should be treated like an exterior job. Jambs and sills should be waterproofed. In this case, tile fills that role. Top left: A block, with one edge already buttered with mortar, is being set onto the mortar bed. The wood strip on the wall acts as a guide to keep the blocks plumb. Top right: Additional rows of glass block are buttered on one edge and pressed into a layer of fresh mortar. In this case white thinset was smeared on the wall to increase the bond between the mortar and the wall. Once the blocks are in place and the joints have been tooled, any mortar haze should be removed with a sponge and plenty of clean water, as shown at right. Ceramic tile caps the edges of the blocks. The drawing below shows typical installation details.

Installing glass block

Other types of panel anchors

Panel anchors: Fasten to structure and embed in mortar

Fiberglass expansion strip for large panel installations

Panel reinforcing (wire ladder)

Mortar

8-in. glass block

Extra cripple studs for support

Caulk

trickier. The panel must be isolated from the surrounding framing to allow for expansion, contraction or deflection of the framing members. But at the same time, it must be anchored securely to the building. To satisfy both conditions, various kinds of expansion strips and special metal panel anchors can be used to anchor panels to a building (drawing, facing page). One is nailed into the wall and the other is embedded in the mortar.

The use of expansion strips at jambs and headers is essential because they allow the surrounding framing to expand and contract without destroying the panel. The block itself won't move much unless the panel is very large; glass block's thermal expansion rate is .0000047 in. per degree Fahrenheit. Expansion strips are usually 4⅛ in. wide, ⅜ in. thick and 2 ft. long, and are made of dense fiberglass or polyethylene. They can be stapled, nailed or glued in place.

Where the edges of the expansion strips are exposed, fill the gap between wall and panel with packing. A common type of packing, called backer rod, looks something like a rope of stiff foam, and is just pushed into the joint. I prefer to use packing material that's square or rectangular in cross section, like POLY-VOID (Stegmeier Corp., 750 Garcia Ave., Pittsburg, Calif. 94565). Remember to take the thickness of expansion strips into consideration when sizing openings. A bead of caulk covers the packing.

Large exterior panels should be installed like large interior panels, but with the additional precaution of waterproofing. Because there are a number of variables to consider in the design of any large panel, I'd suggest that you consult with your local distributor, who may in turn contact a technical representative at Pittsburgh-Corning.

Reinforcing—Like masonry walls, walls of glass block require a certain amount of metal reinforcing to resist bending stresses. Pittsburgh-Corning offers panel reinforcing "ladders" that consist of two parallel runs of stiff wire that are separated by cross wires. The reinforcing should be placed in continuous rows every third course in standard block walls, and every other course in Thinline and VISTABRIK walls.

Proper reinforcing in the mortar joint is essential on larger jobs and useful on smaller ones, but not every glass-block outlet carries the full line of accessories. When scheduling problems won't allow time for tracking them down, I rely on an old standby to fortify small installations. In a pinch, 9-ga. galvanized wire can serve as both reinforcing and anchoring on small panels. For large panels, I'd definitely stick with the standard Pittsburgh-Corning reinforcing.

Installing a panel—Not long ago, I did a project that involved several different panels of glass block. The block I used for all three panels was 12x12 VUE, by Pittsburgh-Corning. I'll show you how I did the smallest one, a divider between the shower and the tub, because it illustrates the versatility of glass block (photos facing page). As you can see, a block panel doesn't always have to be entirely enclosed by wall.

Earlier I had installed ceramic tile on a wood framework that wrapped around the shower,

along with two vertical lengths of rebar to stabilize the "open" side of the panel. Later the rebar would be covered with more ceramic tile. Because the sill was already waterproof, I dispensed with what is normally the first step in such a job: giving the sill a thick coating of asphalt emulsion (normally, this should be allowed to dry before any mortar is applied) or flashing it with a suitable waterproof membrane.

I glued expansion strips to the wall jambs with asphalt emulsion, and set a wood guide strip along the wall to keep the blocks running true. Such a guide speeds the work by giving me a constant reference to plumb. Panels should be anchored to the wall just above the first course, just below the last course, and at some intermediate courses (every 24 in. for standard block, and every 16 in. for Thinline). For this panel, L-shaped lengths of 9-ga. wire, fastened to the wall with 10d nails, were sufficient to serve both as panel reinforcing and anchoring.

When I was ready to lay the block, I mixed up a batch of mortar on a mudboard, scooped up a trowelful of mortar and slid it onto the sill, using a sweeping motion to spread it. All layers of mortar (the mortar bed) should be a full thickness, not furrowed. The slick edges of glass block need 100% support on this first course, so they must be bedded completely in the mortar. The first block went against the expansion strip and, as with all the remaining blocks, I gave it a couple of raps on the top with my fist or a rubber mallet to seat it securely. The block should butt up tight against the expansion strip, but should not be mortared here. The next block was buttered with mortar on one edge, then slid against the first and tapped in place. If you were doing a longer row of blocks you'd just continue this routine of butter, tap, butter until you reached the end of the first row. After every few blocks I carefully check for plumb and level, using a small level, and cut away any excess mortar from the blocks with the edge of a trowel.

Reinforcing should be centered in the mortar bed and run the full width of the panel. When I have to use more than one length of reinforcing to reach the end of the row, I overlap the pieces at least 6 in. or so. The panel anchors should be placed directly over the reinforcing, but they don't need to be fastened together; mortar provides the connection between them. I bend the anchors so that 6 in. to 8 in. contacts the jamb.

While stacking the rows of block, be sure to maintain plumb and level. Strings tacked to matching layouts above and below the panel are especially helpful when laying up curved or serpentine panels. The important thing to remember about glass-block walls is that light will be passing through the finished panels to highlight any inconsistencies in the mortar joints. Very slight variations in joint thickness are acceptable, but goofs or sloppy work practically scream for attention.

It's sometimes hard to maintain a consistent mortar-joint thickness, particularly if the installation is a tricky one, or if you're a novice. One trick is to use wood spacers cut from scrap stock. Wiggle them into the mortar bed just before setting a course of block, and they'll support the blocks while the mortar sets up. Soak

Sources of glass block

Only one company in the U. S. manufactures glass block on a commercial scale:

Pittsburgh-Corning Corporation
800 Presque Isle Dr.
Pittsburgh, Pa. 15239.

Other sources of glass-block:
Forms & Surfaces
Box 5215
Santa Barbara, Calif. 93108
(Japanese glass-block)

Euroglass Corp.
123 Main St., Suite 920
White Plains, N. Y. 10601
(French glass-block).

Two companies import glass block from West German manufacturers:
Solaris U. S. A.
Division of Sholton Associates
6915 S. W. 57th Ave.
Coral Gables, Fla. 33143

Glasshaus, Inc.
P.O. Box 517
Elk Grove Village, Ill. 60007.

the spacers before you put them in so they don't suck moisture from the mortar, and stuff the holes that remain with mortar.

Finishing the job—When the mortar has stiffened a bit, the joints should be tooled. On a large project you can't wait until all the block is up before tooling the joints, but usually you can on a small panel like this one. When the last row is in place, I step back and look at the joints. I usually find voids that need a little extra mortar, so I fill them in. When the voids are filled, each joint can be smoothed with a striking tool, a metal bar with a C-shaped cross section (available from masonry-supply outlets). This forces the mortar into any voids between the blocks, and also it compacts it to make it harder and more waterproof. If you can't get a striking tool, you can use a length of smooth copper pipe.

After tooling the joints on both sides of the panel, I clean up each block with a wet brush or a sponge, rinsing it frequently. If you have ever grouted tile, you will undoubtedly find this step familiar. Any lingering light haze can be removed with a piece of cheesecloth.

On some jobs, to add a little color to the transparent wall, I'll rake out the joints to a depth of ¼ in. to ⅜ in. and fill the resulting channel with colored grout, which I then finish smooth with a striking tool and clean as above.

Glass blocks can be tricky to install properly, but they can also be loads of fun. Just wait for a winter's day when the snow is piled up around the house and the sun is low in the sky. That's the time to pull out the old lawn chair, stretch out and close your eyes—you're as good as on the beach in Jamaica. □

Michael Byrne is a tilesetter in South Hero, Vt. Photos by the author, except where noted. His book, Setting Tile, *was published by The Taunton Press in 1995.*

Acrylic Glazing

How and where to use this plastic instead of glass

by Elizabeth Holland

Commonplace now as the stuff of automobile lights, bank security windows, gas-station signs, camera and contact lenses, TV screens, and even paint, blankets and carpets, acrylic plastic has been around a long time. Although development of this highly elastic substance began back in the 19th century, it wasn't until the 1930s that chemical firms first began producing commercial quantities of acrylic, which can be manufactured as a liquid, as fibers or in sheets. And it took World War II, when the War Department started testing and using acrylic extensively in aircraft, to push the technology into the applications familiar to us today.

The larger family of plastic glazing materials has been closely scrutinized over the last decade by solar designers and builders searching for the least expensive material for collectors, greenhouses, windows, skylights and water storage. Most plastic glazings are flexible, lightweight, impact-resistant and light-diffusing. Acrylic stands out because it will not degrade or yellow in ultraviolet light. Along with high clarity and an impact resistance of 15 to 30 times greater than that of glass, acrylic offers a lifetime gauged at 20 years.

Given a burning rate of Class II in the codes, acrylic burns very rapidly, but does not smoke or produce gases more toxic than those produced by wood or paper. The ignition temperature is higher than that of most woods, but acrylic begins to soften above 160°F.

Used for exterior and interior windows, doors, skylights, clerestories and greenhouse glazing, acrylic sheets can be molded into various shapes and contours, as shown in the photos at right. Both single-skin and double-skin versions of the material are available. Single-skin acrylic is clear and comes in sheets or continuous rolls of various thicknesses. Extruded into a hollow-walled sheet material, double-skin acrylic has interior ribs, spaced $\frac{5}{8}$ in. apart, running the length of the sheet. It is translucent, but not transparent.

The debate—Builders and designers who have worked with acrylic fall into two camps: they either hate it or love it. Any type of glazing is ultimately compared with glass, and those who like acrylic, whether single or double skin, offer these reasons:

It's versatile. Single-skin acrylic can be cut into a multiplicity of shapes, either for pure design reasons or to meet the demands of an out-of-square solar retrofit. Single-skin acrylic can

Photos: Valerie Walsh

Acrylic is light, easy to cut, and has more impact resistance than glass, though it scratches easily and has a high rate of expansion. Double-skin acrylic sheets are translucent, and the diffuse light is good for plants. Designer Valerie Walsh thermoforms these sheets into curved roof sections for custom sunspaces.

be cold-formed into curves; both single and double-skin sheets can be heat-formed.

Acrylics are easy to cut and can be site-fabricated. Lighter in weight and easier to carry, the double-skin sheet is more convenient to install than glass. The flexibility of single-skin sheets varies with their thickness; longer sheets of thinner acrylic require more people to handle them.

Acrylic has high transmissivity, and better impact strength than glass. Double-skin acrylic has an R-value competitive with insulated glass units. It is safe in overhead applications, because it will not shatter. Instead it breaks into large, dull-edged pieces.

On the other hand, builders who prefer glass offer these reasons: Acrylic has a high rate of expansion and contraction, requiring careful attention to keep an installation leak-proof. As it moves, the acrylic sheet makes a noise described as ticking or cracking. And acrylic scratches easily. The extent to which this is considered a problem, or even an annoyance, varies from builder to builder.

Costs—Acrylic used to be much cheaper than glass, but now single-skin acrylic is competitive only when purchased in bulk. Any cost advantage is likely to be lost if you attempt to double-glaze with single-skin acrylic. This is a labor-intensive process, and it's tough to eliminate condensation between the panes. The price of double-skin acrylic is close to that of insulated glass, but the cost is higher if the price of a compression fastening system is figured in. (Some builders expect the price of double-skin acrylic to drop in the future when more companies begin to manufacture it.)

If acrylic is used for a roof in a sunspace or greenhouse, however, its availability in assorted lengths can cut down on labor costs. Designer Larry Lindsey, of the Princeton (N.J.) Energy Group (PEG), points out that long pieces eliminate the need for horizontal mullion breaks, and so can be installed less expensively than several smaller ones. They're also cheaper. An uninterrupted piece of glazing can run the full length of the slope, supported by purlins underneath.

Professional use—Architect David Sellers of Sellers & Co., an architectural firm in Warren, Vt., explains his extensive use of acrylic: "Our whole plastics experiment has been an aesthetic means of expanding the type of architecture we do. With acrylic we could push the house beyond what it was already, both the inside and the outside experience of it." In the process, the firm has developed a spectrum of applications for single-skin acrylic (sidebar, below right).

Designer-builder Valerie Walsh, of Solar Horizon, Santa Fe, N. Mex., uses double-skin acrylic for a portion of the roof in the custom-designed greenhouses and sunspaces that are her firm's specialty. She first used single-skin acrylic because it was slick, clean-looking, and didn't degrade in the Southwestern sun. She began to explore unusual shapes, such as a wheel-spoke roof design. Then she turned to using double-skin acrylic. Walsh thermoforms acrylic in her own shop—curved pieces that are as wide as 5½ ft. and typically 6 ft. to 7 ft. long, although she has done 8-footers.

Safety and economics figured prominently in the Princeton Energy Group's decision to use acrylic glazing overhead in their greenhouses and sunspaces.

"The whole issue is a matter of expense," says Larry Lindsey. "In order to have glass products we feel comfortable installing overhead, we have to pay two penalties, one in transmittance and one in bucks. At present, there is no laminated low-iron glass available at a reasonable cost."

For those who have years of experience with acrylic, a willingness to experiment and to learn from mistakes has produced a valuable body of knowledge about working with the material, its design potential and its limits.

Movement—Leaking is a particular concern with acrylic glazing because it moves a lot, expanding and contracting in response to temperature changes. To avoid leaks, design principle number one is to try to eliminate horizontal joints. And wet glazing systems that may do a perfect job of sealing glass joints will not work at all with acrylic. Its movement will pull the caulk right out.

"Acrylic has a tremendous coefficient of expansion—you have to allow maybe an inch over 14 ft. for movement," cautions Chuck Katzenbach, construction manager at PEG, where they have worked with double-skin acrylic for exterior applications and single-skin for interior ones. "No silicones or sealants we know of will stretch that potential full inch of movement." Indeed, one builder tells a story about using butyl tape for bedding: The acrylic moved so much in the heat that the tapes eventually dangled from the rafters like snakes.

Room for expansion must be left on all four sides of an acrylic sheet, because the material will expand and contract in all directions. The amount of movement depends on the length of the sheet and the temperature extremes it will be subject to.

Acrylic glazing can be installed year round, but it is vital to pay attention to the temperature when it is put in place. Katzenbach explains that if it's 30° outside, then you have to remember to allow for expansion to whatever you figure your high temperature will be. If it could go from 30°F to a peak of 120°F in your greenhouse, you have to make provisions for a

The pliant possibilities of acrylic. For several years now, the architects at Sellers & Co. have been toying with supple single-skin acrylic to carve shapes and sculpt spaces that abandon the simple linear notion of a house. The concepts that have developed, both successful and unsuccessful, are abundant: sliding doors and windows, cylindrical shower stalls, fixed curved windows, curved windows that spring open at the bottom, bus-style fixed windows, skylights, removable windows held in with shock cords, a continuous window up the front of a house and back down its other side, and an entire roof double-glazed with ⅜-in. acrylic.

In the late 1960s, curved windows, shapes that would curl in and out from a house, began to fascinate Sellers. First he tried using a heat lamp to form a 12-in. radius curve in a ⅜-in. acrylic sheet. It worked, but the heat produced some distortions. Then he found he could cold-form the sheet into an absolutely clear curved window, just using the building's structure to hold the sheet in the desired shape (photo, top). The first attempt at curving a piece of ³⁄₁₆-in. acrylic into a shower stall, 8-ft. tall and 6-ft. in diameter, revealed that curving acrylic gave it amazing strength.

From windows, the designers turned to curving continuous sheets of acrylic, some as long as the 45-ft. strips on the sculpture studio at Goddard College, in Plainfield, Vt. The strips stretch from the peak of the roof down to a slow bend at the bottom edge. The limits of curving acrylic became apparent: in a long vertical piece with a curved bottom, the sheet is exposed to incompatible bends, horizontal on the top and then vertical on the bottom, a situation that leads to cracking where the curve begins. In addition, there is the stress of the predominantly vertical movement of such long pieces.

At the Gazley House (photo above), a new detail was tried to support the curve and prevent the cracking. The center bay of the house is glazed with three 23-ft. long strips of single-skin acrylic, fastened with a commercial compression system. Inside and right behind the glazing, a series of 6-ft. tall wooden ribs, similar in appearance to inverted wishbones, support the curve. The carved wooden ribs double as planters. For an extra measure of support, the designers also installed a horizontal metal crosspiece under the acrylic at the point where the curve begins.

The experimentation with acrylic continues: The firm has begun using a strip heater to bend the edges of single-skin acrylic. Installing adjacent strips would require simply bending up the side edges and then capping the adjacent ones after they were in place. Site-fabricated skylights with wrapped edges would be simple to finish and much less expensive than commercial units. The ideas are just beginning to suggest themselves, and designer Jim Sanford thinks the potential could revolutionize what they do. —*E.H.*

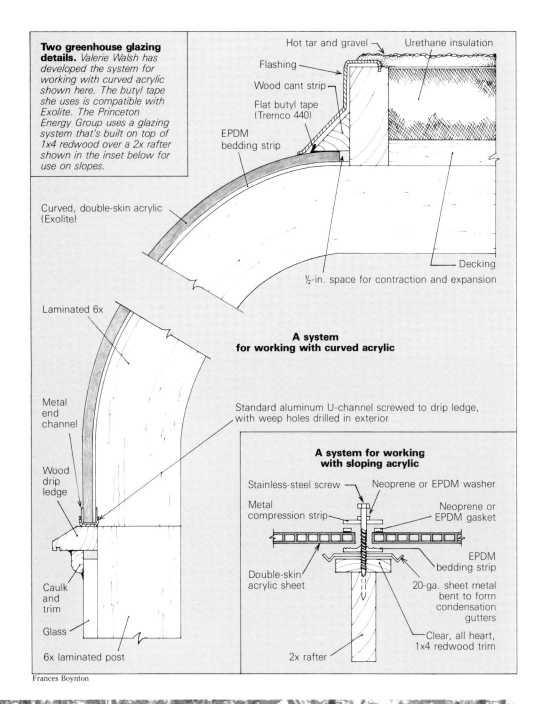

Two greenhouse glazing details. *Valerie Walsh has developed the system for working with curved acrylic shown here. The butyl tape she uses is compatible with Exolite. The Princeton Energy Group uses a glazing system that's built on top of 1x4 redwood over a 2x rafter shown in the inset below for use on slopes.*

Hot tar and gravel

Urethane insulation

Flashing

Wood cant strip

Flat butyl tape (Tremco 440)

EPDM bedding strip

Decking

½-in. space for contraction and expansion

Curved, double-skin acrylic (Exolite)

A system for working with curved acrylic

Laminated 6x

Metal end channel

Standard aluminum U-channel screwed to drip ledge, with weep holes drilled in exterior

A system for working with sloping acrylic

Wood drip ledge

Stainless-steel screw

Neoprene or EPDM washer

Metal compression strip

Neoprene or EPDM gasket

Caulk and trim

Double-skin acrylic sheet

EPDM bedding strip

20-ga. sheet metal bent to form condensation gutters

Glass

6x laminated post

2x rafter

Clear, all heart, 1x4 redwood trim

Frances Boynton

90° change. PEG uses the formula below to calculate the extra space to allow for double-skin expansion:

$$\frac{K}{L} \times \Delta F, \text{ where}$$

K = coefficient of expansion (manufacturer's specs)

L = the length of the glazing (in inches)

ΔF = the difference between the lowest and highest temperatures you expect.

Cutting—To cut a sheet of acrylic, use a fine-tooth carbide-tipped blade set for a shallow cut, and move like a snail. This is important because speed will cause little pressure cracks to appear on the bottom edge. While cutting, make sure that the sheet is firmly supported on both sides of the cut. Sharpness is vital, so use that blade only for working with acrylic. When the acrylic is cut, it heats up and the edges melt, but the wider kerf of a carbide blade will prevent the newly cut edges from melting back together again. After the cut, the edges can be planed, filed or sanded.

As the acrylic is cut, little fuzzy pieces will fly up. Some will reglue themselves to the edges and can be broken off when the cutting is completed. With double-skin acrylic, the fuzz tends to fill the ⅝-in. dia. columns between ribs. Use an air gun to blow it out.

If you drill acrylic, support the sheet fully, and use a very sharp spade bit, ground to a sharper angle than for drilling wood. The sharper angle helps prevent cracking. Drill very slowly, and slow down even more just before the drill breaks through the sheet. Be prepared to break some pieces, no matter how careful you are.

Acrylic sheets come protected with an adhesive masking, which exposure to rain or sunlight makes quite difficult to remove. Leave the protective masking on the acrylic as long as possible, and be prepared for a good zap from static electricity when you pull it off.

Fastening—For years it has been common practice to fasten single-skin acrylic by screwing it down. The designers at Sellers & Co. developed a pressure-plate fastening system to distribute the pressure evenly, and drilled the holes for the bolts or screws an extra ⅛ in. wide to allow for movement. But after ten years or more, the hole has shifted and started pushing against the screw in some installations. Cracks developed where none had existed.

Small cracks in single-skin acrylic can be stopped if the force on them isn't too great. Although his firm now uses installation details that don't involve drilling the sheets, Jim Sanford at Sellers & Co., recommends stopping cracks by drilling a ¼-in. dia. hole at the end of

Acrylic sheets can be cut on site with a sharp, fine-tooth, carbide-tipped blade in a circular saw. Work very slowly to avoid pressure-cracking, then plane, file or sand the new edges smooth. Be sure to support the sheet along both sides of the cut. Photo: Flora LaBriola.

the crack and filling it with silicone. Designer-builder Alex Wade, of Mt. Marion, N.Y., who still uses screws, suggests drilling a tiny hole at the end of the crack, too. He then widens the crack slightly with a knife and fills it with silicone. Finally, he removes the offending screw. Wade suspects that many builders don't take into account the season of the year in which they are working when they drill the holes for screws. When installing acrylic in the extremes of summer or winter, Wade drives the screw either to the inside or the outside of an oversized hole in the acrylic, to allow for subsequent contraction or expansion when the temperatures change. It's important to space the holes evenly (about 2 ft. apart) and to tighten the screws uniformly to distribute the pressure equally. The sheet must be held down firmly, but still be able to move.

To avoid taking a chance with cracking, however, most builders have abandoned screws. Instead they use a compression system of battens that hold the plastic sheet down on a smooth bed of ethylene propylene diene monomer (EPDM). It comes in strips and is supplied by several manufacturers. The acrylic can easily slide across the EPDM as it moves.

Manufacturers recommend a 3x rafter to support the bedding in the compression glazing system. PEG installs an interior condensation gutter on greenhouse rafters that doubles as a smooth, uniform bed for the EPDM gasket in the glazing system. The 20-gauge sheet metal straddles the rafter and is bent into a $\frac{5}{8}$-in. lip for the gutter on each side.

PEG has also developed a system for a standard 2x. A clear all-heart redwood 1x4 trim piece is screwed on top of the 2x, widening the bed and providing a smooth surface (inset drawing, facing page). Concentrated stress on the acrylic sheets is as important to avoid with a compression system as it is with screws. If one point is fastened tighter than the others, the acrylic will bow in and leak or crack.

Larry Lindsey recommends aluminum battens on south-facing roofs, because wooden ones will eventually cup upward, creating a leak. Aluminum battens can be purchased with various finishes, or they can be capped with a strip of redwood.

Double-skin acrylic needs to be supported at its base or it will bow instead of moving within the compression glazing system.

PEG lets the sheets hang over the roof's edges as a shingle would, sealing it underneath. For the bottom edge on installations with curved roofs, Valerie Walsh has developed a system with no damming problems. She slides on aluminum terminal section (ATS) from CYRO (697 Route 46, P.O. Box 1779, Clifton, N.J. 07015) on the bottom of the double-skin acrylic, then snugs the acrylic into a larger aluminum U-channel that is in turn screwed into the wood beam. Walsh then caulks the inside and drills weep holes through the outside of the U-channel (large drawing, facing page).

In one of his designs, David Sellers decided to glaze the south-facing roof area with long strips of single-skin acrylic. To avoid leaking, he encouraged the tendency of the $\frac{1}{4}$-in. sheets to

Single-skin acrylic is flexible, and some designers take advantage of this to encourage drainage. Extra blocking along the sheets encourages a sag that carries water away. Photo: David Sellers.

sag slightly. Small blocks under the edges of the sheets accentuate the dip. Melting snow or rain flows to the center of each panel and then drains off the roof. On the bottom edge, an angle keeps the acrylic from slipping, as shown in the photo above.

Glazing materials—Acrylic is fussy stuff. Chuck Katzenbach reels off a list of materials to avoid using with this plastic. Vinyl leaches into acrylic and weakens its edges. Some butyls have plasticizers that may also leach into acrylic. In these cases, either the acrylic will eventually fail or the butyl will become very hard. The plasticizer in most neoprenes is not compatible, so check with the manufacturer.

Compatible glazing materials are few: EPDM heads the list. Silicone caulk is okay, but sooner or later the acrylic's movement will pull it loose. If it's installed on a cold day, the silicone may pull out on the first really warm one. There are many urethane foams you can use, but you would be well advised to consult the manufacturer directly about compatibility.

Support—The double-skin acrylic can bow in over the length of the roof, and the sheet could conceivably pull out of the glazing system under a very heavy snow load, according to Larry Lindsey. So as a cautionary measure, PEG figures on a 30-lb. snow load and installs purlins 4 ft. o.c., about $\frac{5}{8}$ in. below the sheet.

Condensation—Acrylic transpires water vapor, so a double-skin unit typically will have some cloudy vapor inside. Double-skin acrylic should be installed with its ribs running down the slope so that any condensation inside the channels will collect at the bottom edge of the sheet. This edge needs to be vented. Double-skin sheets arrive with rubber packing material in both ends of the channels, to keep them free of debris. PEG's construction crew just leaves it there and perforates it with a scratch awl to allow air movement.

Cleaning—Never use abrasives, ammonia-base glass cleaners or paint thinner on acrylic glazing. Mild detergent, rubbing alcohol, turpentine and wax-base cleaner-polishers designed especially for plastic are safe when applied with a soft cloth.

Cleaning brings up the controversial issue of scratching. "The scratching drives me crazy, though other builders don't seem to care as much," says Valerie Walsh. Whether she is storing the sheets she has heat-formed into curves or transporting them to the building site, she keeps thin sheets of foam padding wrapped around every piece.

"With double-skin acrylic, people's tendency is not to expect to be able to see through it. They're not looking through it as they would through a window, and so they're not seeing the small imperfections in the surface itself," argues Chuck Katzenbach.

Even though single-skin acrylic is transparent, many builders who work with it say that scratching just isn't a significant problem, particularly if the glazing is kept clean. The only serious scratching problem is likely to be the work of a dog. Most scratches can be easily removed with a Simonize paste-wax buffing. And now some single-skin acrylics are available with a polysilicate coating that makes them abrasion-resistant and also improves their chemical resistance. □

Elizabeth Holland, of West Shokan, N.Y., is a contributing writer to this magazine.

Sources of supply
Acrylite (single), **Exolite** (double): CYRO, 697 Route 46, P.O. Box 1779, Clifton, N.J. 07015.

Lucite (single): DuPont Co., Lucite Sheet Products Group, Wilmington, Del. 19898.

Plexiglas (single): Rohm and Haas, Independence Mall West, Philadelphia, Pa. 19105.

Acrivue A (single, with abrasion-resistant coating): Swedlow Inc., 12122 Western Ave., Garden Grove, Calif. 92645.

Site-Built, Fixed-Glass Skylights

An energy-efficient, watertight design that you can build with standard materials

by Stephen Lasar

Fixed-glass, site-built skylights can provide many cost and thermal benefits for new construction, as well as for additions to existing structures. And having a reliable design that can be built on site gives the architect or builder the flexibility to meet functional and aesthetic requirements. The skylight design that we use most often is based on standard techniques and materials, and any skilled carpenter can install them. So far, our skylights have withstood a wide range of weather conditions without failing. And the prepainted aluminum flashing and battens we use give these skylights a clean, unobtrusive appearance.

The curb—In holding the glass above roof level, the curb keeps runoff and debris off the skylight. The curb is made from straight-grain 3x10 Douglas fir that we rip in half to yield two 3-in. by 4½-in. curb members. We then rabbet them along one edge to accept the glass. Center curbs, if any, hold two glass panels, and are rabbeted along two edges. The depth of the rabbet is 1¼ in., which is ¼ in. less than the combined thickness of the insulated-glass panel with its two glazing strips. We take up this extra ¼ in. by compression when we screw down the battens to form a weathertight seal (drawing, facing page). The rabbet is ¾ in. wide. We size the curb enclosures so that ½ in. of this width holds the glass and glazing tape. This leaves ¼ in. of open float space between the edges of the panel and the wood.

We assemble the curb box on the roof, using butt joints at corners and a housing (dado) to let each center curb into its top and bottom curb. Toenailing the curb to rafters and headers is usually enough to hold it securely, but if the roof pitch is steeper than 12 in 12, then we use metal clips, too. For structural reasons, we use 3x10 or 3x12 rafters on most spans beneath central curbs (photo below left).

Site-built skylights call for flashing that is also cut and formed on site. The system shown on p. 77 works well with this type of curb. The aluminum should be .019 in. or thicker, and at least 8 in. wide. I always specify prepainted flashing stock because it blends in with a new roof. Shiny, unpainted aluminum draws needless attention to the skylight.

Tape and glass—The inside tape we use is Preshim 440 ¼-in. by ½-in. spacer-rod tape made by Tremco, Inc. (10701 Shaker Blvd., Cleveland, Ohio 44104). It's a butyl-base material and is compatible with the silicone second seal of the insulated-glass units we use (be sure to check caulk-sealant compatibility very carefully). The spacer rod, or shim, is continuous, and limits the compressibility of the tape, so that the glass won't settle away from its seal over time. It also helps to prevent tape squeeze-out

at pressure points. Sticky on both sides, the tape comes in rolls, with one side faced with paper. It is applied to the bottom of the curb's rabbet, set slightly back from the vertical edge. It's important to lay the tape carefully, so that it has room to expand without touching the inside of the rabbet or protruding over its outer edge. Butt each tape so there are no airspaces between sections, and lay the tape with the paper facing up (photo below). Don't strip off the paper yet, because you may have to shift the glass panel after you lay it down.

The glass is installed from the outside, so be sure you rig a secure scaffold. The insulated panels used here are standard, double-pane tempered sliding-door units, 92 in. long, 34 in. wide and 1 in. thick. Don't try to lift and position the glass by yourself; get at least one helper. You won't need suction cups for getting the glass onto the roof, but you will need them to lower the panels down into the curb and to shift them into final position.

Before setting the glazing in place, put a neoprene setting block (¼ in. thick, 4 in. long and 1 in. wide) one-quarter of the way in from each bottom corner of the rabbeted curb. These spacers hold the panel away from the bottom edge, giving the glass room for expansion. Now you can set the bottom edge of the panel against the setting blocks and lower it carefully into the curb. Center the panel and make cer-

With the rabbeted curbs nailed to rafters and flashed to the roof, the next step, above, is to lay the glazing tape. The paper should be left on the tape until the glass panels are centered. Then the top layer of glazing tape is applied, right. A batten joint will hide the flashing and tape that cover the curb between panels. Tape is laid against the edge of the glass to allow ¼ in. of float space on all sides.

tain that you've got your ¼ in. of float space around all four edges. To do this, you'll have to pull up the panel at the top, adjust it, and set it down again. Once the position is right, pull up the panel one last time, strip the paper facing off the glazing tape and set the glass panel down for good.

The next step is to apply the outer glazing tape. We use Tremco Polyshim tape, a butyl-base, compression-type tape with good adhesion and elastic qualities. As with the tape beneath the glass, this glazing strip shouldn't butt right against the wood. Leave about ¼ in. of space between wood and tape for expansion. Once you've taped the glass, you should also flash and tape all corners of the curb as shown in the small photo on the facing page. Wherever aluminum battens will intersect, cover the curb with a strip of flashing and a length of tape. Remove the shim from the tape so that when the battens are screwed down, the glazing tape will be squeezed into the joint to form a weathertight seal.

For battens over the curb's perimeter we use 3-in. by 3-in. by ⅛-in. pre-painted aluminum angles. We cut the battens for center curbs from flat aluminum stock 3 in. wide and ¼ in. thick. The side angles should be cut with ears at both ends to create an interlocking joint with top and bottom battens, as shown in the drawing at right. All top and side pieces also need to

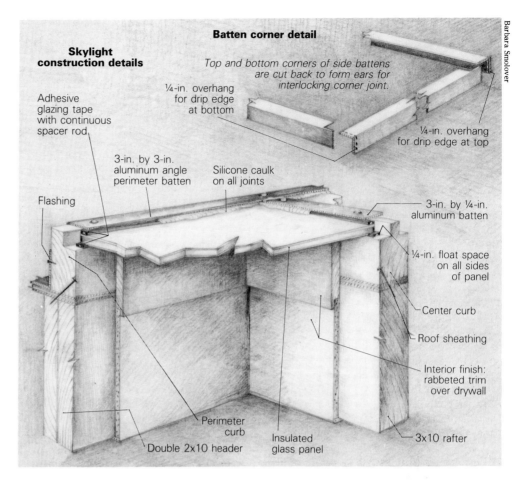

Batten corner detail

Skylight construction details

Top and bottom corners of side battens are cut back to form ears for interlocking corner joint.

¼-in. overhang for drip edge at bottom

¼-in. overhang for drip edge at top

Adhesive glazing tape with continuous spacer rod

3-in. by 3-in. aluminum angle perimeter batten

Silicone caulk on all joints

Flashing

3-in. by ¼-in. aluminum batten

¼-in. float space on all sides of panel

Center curb

Roof sheathing

Interior finish: rabbeted trim over drywall

Perimeter curb

Double 2x10 header

Insulated glass panel

3x10 rafter

Below, a flat aluminum batten is laid down over a central curb. It has been predrilled for screws, and will cover the top layer of glazing tape at the edge of the glass panel. Details of the skylight construction are shown in the drawing above.

Top, stainless-steel screws with neoprene and stainless-steel washers pull the battens tight, compressing the glazing tape for a weathertight seal. Above, the last step in building the skylight is to caulk all batten and glass-to-batten joints. For best results, warm the caulk before you start to apply it and keep the bead continuous.

be cut slightly long to create drip edges. The perimeter battens cover the curb flashing. Install them first, over a generous bead of silicone caulk. Where battens meet, leave about 1/16 in. between metal edges, for caulking and for the glazing tape to squeeze into. We use 1 1/2-in. long stainless-steel No. 10 Phillips panhead screws with stainless-steel and neoprene washers to pull the battens down tight over the glazing. As for screw spacing, 16 in. o.c. is good for angled battens; 12 in. o.c. for flats.

Replacing a defective panel of insulated glass is an irritating and expensive job. Not considering breakage, failure of the seal or spacer between panes is the most common problem with insulated-glass panels. It's a good idea to specify a double seal that has a good rating from the Insulating Glass Certification Council. This group publishes "Certified Product Directory," a 28-page booklet that lists the names, addresses and phone numbers of over 50 American manufacturers of insulated glass, along with details on their products' corner construction, spacer materials, dessicants and

sealants. Though intended mainly for manufacturers, it may be of use to architects, contractors and builders. It costs $1 from the IGCC (Industrial Park, Route 11, Cortland, N.Y. 13045).

Caulking is the last step. The joint between the glass and the aluminum batten and the joints between battens should be sealed with an even, continuous bead of exterior-grade silicone, carefully lapped at corners. For best application, get the caulk warm before you start using it. Fill any voids or uneven spaces, and smooth the caulk with your finger where necessary.

There are several options for finish trim on the interior of the skylight. We use either veneered plywood nailed directly to rafters and headers, or a trim strip, rabbeted to cover plaster or drywall. In either case, it's important not to butt the plywood or trim against the glass surface. Allow 1/4 in. to 3/8 in. all the way around the skylight so the glass and wood have room to move. □

Solar architect Stephen Lasar practices in New Milford, Conn.

Flashing a curb
by Jim Picton

Most manufactured skylights can be bought with flashing kits, but if you're installing a site-built unit like the one described in the previous article, cutting and forming your own flashing is the only way to go. Even if you're working with a factory-built unit, you may choose not to buy the optional flashing kit, but to make your own instead. This is an especially attractive alternative when the manufacturer supplies strip flashing for the unit's sides instead of the superior step flashing. Buy a roll of .019-in. thick aluminum 16 in. or 20 in. wide, and follow the directions below.

Base flashing—First unroll a length of aluminum, and cut off a piece about 8 in. longer than the width of the skylight. Slice it to a width of 8 in., and save the scrap for step flashing. Bend the aluminum lengthwise along a straightedge to form a right angle, with one side about 3 in. across and the other, 5 in. Then set the 3-in. side against the bottom curb of the skylight, and the 5-in. side flat on the roof. Hold the angled aluminum so that an equal amount of flashing extends beyond each side of the skylight. Now mark a line on the vertical side of the flashing flush with the corner of the curb on each side of the skylight. This is a fold line. Next, draw a line to make a diagonal cut across the vertical side of the flashing, ending exactly at the point where the fold line meets the bent corner of the flashing, as shown in step 1 of the drawing on the facing page.

When you've cut both sides, fold the vertical half of the flashing around the skylight, and flatten the rest onto the roof. If you're not re-shingling, slide the angled tab of flashing under the first course of shingles on each side of the skylight. This is your base flashing. Folded and cut, it's ready to tack to the roof with a single nail in each upper corner.

Step flashing—Most manufactured flashings include the first two pieces of step flashing as an integral part of the bottom flashing. Each piece is bonded to the bottom flashing with a soldered or locked seam joint. You can approximate this detail using an 8-in. square of aluminum sheet from your roll, bending it at a right angle with 3 in. vertical and 5 in. flat, and making a V-cutout along its fold so that it can be bent around the base (step 2).

The rest of the step flashing, except for the two top corner pieces, consists of square or rectangular sheets with a single 90° bend. The precut sheets sold in hardware stores as step flashing are typically 5 in. by 7 in., a size that offers minimal protection. Buy larger pieces if you can, or make your own from 8-in. squares.

If you're putting on a new roof, lay the next course of full shingles when the first piece of step flashing is in place. Then mark the shingle that will fit next to the skylight and cut it, leaving about 3/4 in. of space between curb and shingle edge. Position the shingle and nail it down at the top edge farthest from the skylight. Now slide the second piece of step flashing under the shingle adjacent to the skylight curb, so that the bottom edge of the flashing is about 1/2 in. higher than the bottom edge of the shingle. The flashing should lie directly under the bottom half of the upper shingle, and over the entire concealed upper half of the lower shingle (step 3). If you're step-flashing a roof that's already been shingled, be sure to lace each piece of flashing over and under the shingles before nailing its upper corner to the roof.

Head flashing—The last pieces of step flashing will be cut and bent around the top corners of the curb (step 4), just like the step flashing at the lower corners. Once you've done this, the head flashing can go on. Bend, cut and fit it just as you did the base flashing, but use a full-width piece (at least 16 in.) from your roll. Before nailing it to the roof, nail one thickness of shingles to the bare roof above the curb. This will shim the roof up to the level of the shingles on each side, and prevent the head flashing from sagging in the middle and collecting water and debris. Remember that the upper part of the sheet must go under the first course of shingles above the curb, while the edges rest on top of the corner step flashing (step 5).

An exception to this procedure is when you are using a manufactured skylight such as Roto Stella, which requires installing the top and bottom flashings at the time the unit is installed. In this case, you have to slide the last piece of step flashing up under the preformed corner of the top flashing, and forget about bending the step flashing around the top.

Manufactured flashings usually won't extend too far up under the shingles, so they often have a bent return on the upper edge of the top flashing. In case water manages to back up under the shingles, this feature is intended to prevent the moisture from traveling beyond the top of the flashing and getting back under the roof.

Site-constructed flashings that are run 16 in. or 17 in. under the shingles do not need a bent return, but you can make one as an added precaution. Using a straightedge and your fingers, bend the top of the flashing up, about ½ in. from the edge. Fold it all the way over, then place a board over it and hammer it flat. Finally, use a screwdriver to pry the bent edge back up a little, so a space is visible between the top edge of the bent return and the rest of the flashing below.

The most difficult part of fitting homemade flashing is working around the four corners of the skylight curb. The points of these corners are the one area in a flashing job where, unless you are using soldered, welded or otherwise pre-formed components, no overlap is possible. The way to solve this problem is to do what carpenters and roofers always do when faced with the impossible: caulk it.

Counterflashing—Sometimes called angled or perimeter battens (drawing, p. 75), counterflashing is a piece of metal (usually a heavier gauge than roof flashing) bent lengthwise at right angles. Half of this flashing covers the top of the curb; the other half protects the sides of the curb and covers the top edges of the roof flashing. On some prefabricated skylights, counterflashing is permanently affixed to the curb, and the step flashing has to be slipped under and worked into position as the roof is shingled. Other units have a removable counterflashing that is attached quite simply with screws once the roof flashing is complete. In either case, it completes the exterior seal. □

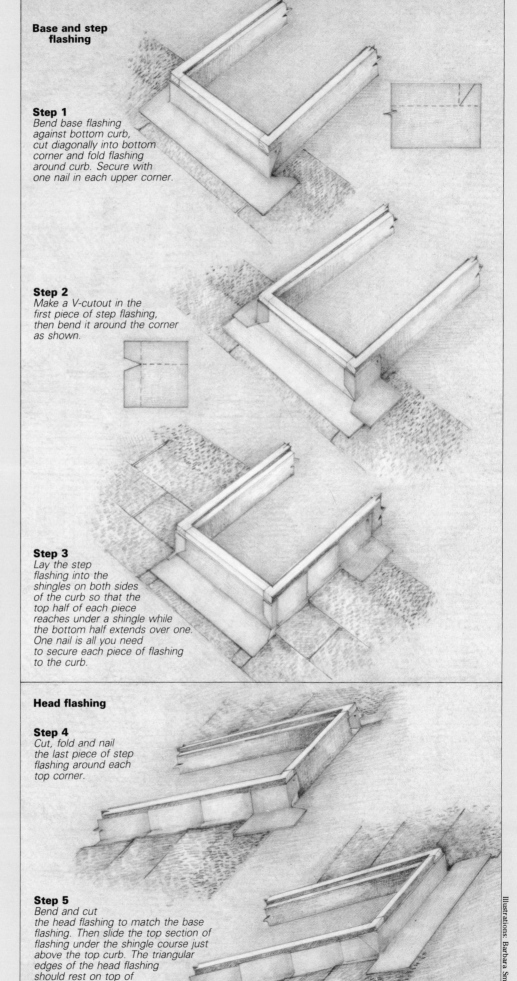

Base and step flashing

Step 1
Bend base flashing against bottom curb, cut diagonally into bottom corner and fold flashing around curb. Secure with one nail in each upper corner.

Step 2
Make a V-cutout in the first piece of step flashing, then bend it around the corner as shown.

Step 3
Lay the step flashing into the shingles on both sides of the curb so that the top half of each piece reaches under a shingle while the bottom half extends over one. One nail is all you need to secure each piece of flashing to the curb.

Head flashing

Step 4
Cut, fold and nail the last piece of step flashing around each top corner.

Step 5
Bend and cut the head flashing to match the base flashing. Then slide the top section of flashing under the shingle course just above the top curb. The triangular edges of the head flashing should rest on top of the step flashing.

Illustrations: Barbara Smolover

Movable Insulation for Skylights

High insulation value and ease of operation
ensure maximum energy returns from your skylights

by Larry Medinger

Skylights, along with properly sized south-facing glass, take best advantage of the sun in climates where winters are mostly cloudy—Oregon, in this case. But finding a convenient and effective way to keep sunspaces from overheating in the summer and losing heat in the winter is a problem. The solution shown here is to mount insulated panels in tracks so that they can fully or partially block the skylight openings. Moving the panels is just a matter of a slight tug on a rope.

Skylights are an asset to any home, particularly in cloudy winter climates. They can be the major source of solar heat gain during most weather, and they can make any room feel light and airy on the gloomiest of days.

Skylights will collect more direct sunlight in months when vertical windows will collect less, and the two glazing systems can reinforce one another to provide a balanced influx of sunlight all year long. Skylights will allow sunlight to penetrate deep into the structure, where it is less expensive and intrusive to the home's design to place thermal mass.

There are many areas in the United States where cloudy winter weather predominates over clear sunny days. In the typical textbook scenario, the low-angle winter sun shines brightly in under the 2-ft. eave overhang and charges the thermal mass. The reality is that over much of the country, for much of the winter, solar radiation comes essentially vertically from the cloud cover overhead.

South-facing vertical glass under a 2-ft. overhang will collect little of the available radiation on cloudy days. For the purposes of daylighting in such conditions, skylights work better than south-facing glass, and they can help collect more of the spare solar heat on these days.

The challenge is to create a comprehensive, climate-sensitive glazing scheme that will be properly sized with the heating and daylighting needs of the house. It must be correctly installed and flashed, and have an insulation system to protect against heat loss on winter nights and to keep the spaces below from overheating on summer days. To provide this kind of movable insulation for skylights, we developed a way to build easily operated, high R-factor, insulating panels for "sky-facing windows."

A first attempt—A number of the homes we've built in the Pacific Northwest use banks of skylights in one or more centrally located collection rooms. Each skylight is mounted between two beams on the ceiling. Our first design featured sliding panels made of rigid-foam insulation in wood frames. The sliding panels were supported between the beams by projecting wooden tracks. The panels could be raised and lowered along the ceiling by a counterweighted system of ropes and pulleys. The contraption worked, but the ropes and wood-to-wood contact of panel and track produced too much friction and made the panels difficult to raise. What we needed was an insulated panel that could be moved easily into place and back out of the way, and one that we could make from off-the-shelf parts.

New tracks—For the tracking system, we finally settled on standard bypass sliding closet-door hardware. We used the wheels without modification. But bypass steel tracks (we used Cox #12-100, which are readily available in hardware stores) accommodate a pair of doors in the same opening. We needed to divide them into two single tracks. Using an abrasive cutoff wheel, we ripped the track into two pieces on our table saw. This operation demands wearing a face shield, as lots of shrapnel from the cutting wheel and the track gets hurled during the cut.

The object is to rip the track exactly in the right place (detail drawing, bottom of page). To keep the track from wandering away from the fence during the cut, you need to clamp a board to the saw table ahead of the blade to hold the track firmly and consistently against the fence. Next, grind or file the sharp edges and dings off the sawn edges to make them reasonably smooth. Sometimes in the ripping process, the tracks take on a slight curve. Be sure to straighten them before proceeding. Then drill ⅛-in. dia. holes in the tracks every 8 in. to 12 in. to accept Phillips-head 1¼-in. drywall screws. The tracks and other hardware are then cleaned with vinegar and spray-painted flat black.

Another problem we had with our first wooden track system was that it was difficult to achieve a durable seal between the ceiling and the panel in its closed position. A good seal is important because it reduces air movement, and thus temperature exchange, between the chilled skylight well and the warm room. It also inhibits condensation on the inner surface of the skylight. While the ends of the panels can be sealed by being made to contact weatherstripped bumpers, the sides have to slide past the surfaces they should be sealing tightly against. This causes friction and wear.

We found a workable solution by mounting the bypass tracks on the beams at a slant to the ceiling, as shown in the photo and drawing above right. The panels themselves are actually kept parallel to the ceiling by mounting the front and rear wheels at different attitudes. This allows the panel to move freely until the moment that it contacts the bottom edges of the skylight well. The pressure of the counterweight or the weight of the panel (depending on whether its closed position is uphill or downhill) keeps a positive contact between the well and the panel, and so maintains a good seal between the two. We have used either commonly available foam weatherstrip gaskets or light-colored fabric rolled under on the edges and stapled to the tops of the panels.

Making the panels—We build our panel frames of clear vertical-grain Douglas fir. For the panel itself, we usually use Masonite and glue it into grooves in the surrounding frame. We expose the smooth side to the room and paint it the color of the ceiling or walls. The frame is finished the same as the nearby beams and the rest of the trim in the room. The insulation in-

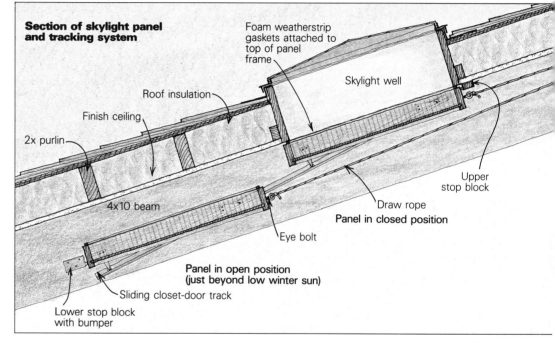

Section of skylight panel and tracking system

Foam weatherstrip gaskets attached to top of panel frame

Skylight well

Roof insulation

Finish ceiling

2x purlin

4x10 beam

Upper stop block

Draw rope

Panel in closed position

Eye bolt

Panel in open position (just beyond low winter sun)

Sliding closet-door track

Lower stop block with bumper

The insulated panels consist of a 1x3 fir frame that's joined with glued dadoes (detail photo at right) and grooved to accept a Masonite bottom. Inside the frame is a tight-fitting panel of 2-in. foil-faced rigid insulation board, with the foil side up. The panels ride in tracks (photo at top) cut from standard sliding closet-door hardware, and the wheels are mounted in a way that keeps the panel from making contact with the ceiling until it's directly under the skylight, where its gasket makes a positive seal. The rope-and-pulley system allows all the panels to be controlled from a single location.

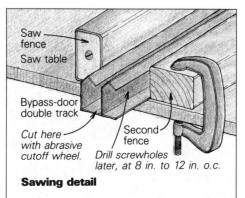

Saw fence

Saw table

Bypass-door double track

Cut here with abrasive cutoff wheel.

Second fence

Drill screwholes later, at 8 in. to 12 in. o.c.

Sawing detail

Photos above and facing page: Jonathan Landis; Drawings: Vince Babak

Counterweights with tubular-sock covers are sized to match the weight of the shutters. An easy tug on the cord and the shutters can be moved up, and they will stay put in any position.

side the panel can be any good-quality 2-in. foil-faced rigid foam. The foam should be fitted tightly into the frame. We use a panel adhesive that is compatible with the foam (check with the manufacturer of the foam) and glue it to the top of the Masonite to help eliminate warping. Make your panel frame a little deeper than the nominal 2-in. thickness of the insulation. You will often find that it is a little thicker.

Installation tips—Before mounting the tracks on the beam sides, screw the wheels on the panel frame. As the drawing on the previous page shows, they should be set at angles that will position the ends of the track within an inch of the top and bottom edges of the beam. The wheel-mounting angles are determined by the length of the panel and the depth of the beam. Laying this out schematically full scale on a plywood sheet can save a lot of tedious test-fitting.

Next, get a helper to support the panel against the ceiling in the closed position under the skylight well. The tracks can be slid into position over the wheels on each side of the panel and screwed in place. We use 1¼-in. drywall screws to fasten the tracks to the sides of the beams. Hold the track tightly against the bottom (the running surface) of the wheels until at least two screws are in place in each.

If we find that one or more of our beams has warped in curing, we make the panel to accommodate the smallest width in its run. Then we shim out the tracks in the wider portions of the run so they're exactly parallel to each other. For shims, we use either small precut pieces of stiff cardboard or thin washers sprayed flat black.

Once you have the exposed part of your tracks screwed to the beams, move the panel to that part of the track so that the rest of the track can be fastened. You may find upon moving the panel back and forth that you will want to adjust the shimming slightly. Sometimes this is just a matter of screw pressure.

When you are satisfied with your track operation, install two braking blocks on the downhill end of the run. These stop blocks should be of the same material as the panel frame and finished similarly. We also fit them with rubber bumper buttons, which are available in most hardware stores. If your panel is in the closed position on the downhill end of the track, adjust the stop blocks' position so the panel can come into full contact with the ceiling. The blocks' primary function is to stop the panel's downhill movement without damage, should someone let it down too fast.

Stop blocks should be installed at the upper end of the run, too. These will keep the panel from being drawn off the end of the track.

Tying things together—The draw rope is tied to an eyebolt set in the center of the frame facing the upward direction of the run. There is considerable stress placed on this piece and so

when assembling the frames we glue the dadoed joints together, as shown in the photo at bottom right on the previous page.

When devising our rope layout, we gather the rope ends to the most convenient location for owner operation. The primary source of friction is in the pulleys used to make the rope turn corners. So buy good-quality pulleys and use the largest size that is practical for your design.

Once the pulleys and ropes are in place, all that's left to do is attach the counterweights to the ends of the ropes so that the panels can be raised and lowered without strain. We tie a temporary loop in the end of each rope and hook a small spring scale to each one in turn. When we pull on the scale, we find two weight levels indicated. The first level is the force needed to stop the downward progress of the panel on its track. The second level is the force needed to begin moving the panel up its track. The range between the two represents the amount of force needed to overcome the friction in the system. We size our counterweight to be within this range on each panel. This allows the panel to be set partially open if desired.

We use 1½-in. round steel stock for our weights, and cut it to the length that will give us the appropriate weight. Finally we drill a hole in the end for the rope and tie it up. As a finishing touch, we cover the counterweights with tubular socks (photo left).

Using the system—This panel design will give you approximately R-16 nighttime insulation over your skylights. It is simpler to devise a high R-value insulating cover for a skylight than it is for a vertical south-facing window because panel storage is not a problem for skylights. Convenience and ease of operation ensure that the owners will be more likely to use the system, and therefore that it will be more effective in saving energy.

In the summertime, the panels may be closed during the hottest part of the day. The foil covering on the insulation may be left exposed to reflect as much light as possible. If the panel tops are visible in their closed position from outside the house, a likely compromise would be to paint the upper foil surfaces white or cover them with light-colored fabric. The sides of the well should always be painted a light color for proper light transmission.

Some summertime heat, however, is bound to be trapped in the well and accumulate to a fairly high temperature. If you glaze your skylights with Plexiglas, you will want to protect it from temperature buildups. One way to do this is to have a seasonally adjusted blocking device to allow your panels to remain just slightly open so that a small amount of convection will moderate any high temperature buildups. Another solution is to build some simple, easily handled lattice boxes that are installed seasonally over the skylight and fastened with wing nuts to hanger bolts set in the skylight curb. With the lattice in place, one may still benefit from the daylighting effect on all but the warmest of days. □

Larry Medinger, of Ashland, Ore., designs and builds energy-efficient passive-solar houses.

Installing a Factory-Built Skylight
Careful selection and minor structural modifications make for a good job

by Jim Picton

Many home owners and craftsmen who are willing to cut into floors and walls will steer clear of retrofitting skylights because of their reputation for developing leaks. This notoriety is largely a result of the failure of older or improperly flashed units, which rely entirely on a chemical seal like asphalt roof cement to keep water out. One such variety, still available today for about $25, is the plastic bubble—a sheet of acrylic with a bulge in the middle—which is plopped down in a bed of cement, with shingles laid over the edges. Successive layers of cement are then applied to the edge surfaces as often as necessary.

We can thank the Arab oil embargo for improving the standards of skylight construction. Interest in alternative energy sources has brought solar heating to the fore, and with it the need for reliable roof windows. While there are still times and places when you will want to build your own units, competition in the marketplace has produced a number of well-designed skylights that are as weatherproof and problem-free as conventional vertical windows. You should still shop carefully, though. Prices vary widely, and are generally an indication of quality.

There are several things to look for when you're shopping for a unit. First, decide whether you want a fixed or operable skylight. Second, consider the flashing package. Step flashing will last the life of the roof. Strip flashing won't. Third, see if the skylight has a thermal break. On some, metal extends from the outside surface to the inside. This can cause a lot of condensation trouble in a cold climate. Fourth, check to see if you need tempered glass. You can save $50 or so by not using it, but building codes specify it where standard glass might easily be broken. Finally, examine the screen setup if you're buying an operable unit. Some are easy to use. Others are a bit quirky: For example, my Velux screen opens and closes with Velcro strips.

Most skylights, whether site-built or manufactured, have curbs 4 in. to 6 in. high, which raise the glazing above the level of the shingles and divert water around the unit. The glazing is spared the cascade that develops as water from the rest of the roof flows to the gutters. In addition, curbs keep granules from the shingles and other debris off the glass. This prevents scratching, and prolongs the life of the glazing seal.

A fixed-glass unit will run between $200 and about $600, depending on flashing and other

Picton's operable skylight was installed between reinforced rafters in a room with a ceiling that followed the roof pitch.

materials. You can get operable skylights for as little as $200, but most cost a great deal more. Crank, spring-loaded and center-pinned units are available, some with strip flashing and some with step flashing. With my contractor's discount, my center-pinned Velux and its flashing package cost me $500.

Structural considerations—If you have a flat ceiling with an attic above, you'll have to build a shaft between the roof and the ceiling. When roof and ceiling are separated only by rafters, as in the installation shown above, the job is more straightforward. In either case, before cutting a hole in your roof, think about the effect of cutting through one or more rafters. As a start, consider how roof openings are framed in new buildings. Double framing is conventional wherever rafters or headers are used to support additional roof-framing members. Rafters and headers that frame the opening are doubled. Cripple rafters run from the headers to the top and bottom of the roof.

Installing double headers is hardly more trouble than installing single ones, but doubling the rafters can be a problem. To be most effective, the double rafters should extend all the way to the points of support—usually to the ridge above and to the double top plate of the wall below.

You can usually do most of your cutting and framing from inside without disturbing the exterior roof surface. This lets you work without worrying about the weather. In some installations though, the whole point is to disturb the interior finish as little as possible, and you may decide to remove some of the sheathing from the outside of the roof, and install double rafters from above. This involves additional re-roofing, but may save you the grief of having to

live in the dust and rubble caused by tearing up the ceiling.

The added load taken by the rafters on each side of an opening can also be offset by reducing their span, or the distance between points of support. If there is a small attic crawl space above the opening, a purlin can be snugged into the area between the collar ties and the rafters. Although collar ties themselves reinforce the rafters, a purlin will cut down the span of the rafters and transfer the load to the walls.

Another alternative is to double up only a portion of the side rafters. A piece of lumber the same thickness and width as the rafter, extending a few feet above and below the skylight opening, and nailed solidly to the rafter, can have a stiffening effect.

When you're selecting a skylight, consider the width of the unit in relation to the rafter spacing. Choosing the next smaller size may mean cutting through one less rafter. Get the advice of a professional, a structural engineer or the local building inspector, if the problems you foresee do not have direct solutions.

Roughing out inside—Once the location for the skylight is selected, mark the rough opening for the unit on the ceiling. It should be about ¼ in. larger all around than the outside dimensions of the skylight unit you've settled on. Measure up 3 in. from the top line you've marked, and down 3 in. from the bottom one. You'll need this extra space for the new double headers. Next, find the rafters at either side of the opening and mark cut-lines that follow their centers. When the opening is completed, the ceiling will be patched at these lines, and the rafters will provide a nailing surface for drywall or finish trim.

Cut the ceiling using a utility knife, or a skill-saw set to cut only the thickness of the ceiling material. You don't want to cut through hidden wiring. When the ceiling panel is removed, pull out any exposed insulation and re-route the wiring if necessary.

To prevent the roof from sagging slightly when the rafters are cut through, support the rafters above and below the opening with temporary braces made by knocking together two 2x4s in the shape of a T, and wedging them between the rafters and ceiling joists. Then cut the rafters to be removed square, 3 in. back from the rough opening line.

Install the double headers one at a time. Cut the first piece to fit between the side rafters,

Picton removes a rafter, above, having already reinforced those on either side by doubling them up. The T-brace to his left keeps the roof from sagging before headers are installed.

Both boards of the double header are cut to fit between the side rafters, left, then installed one at a time. They are toenailed between the side rafters, then face-nailed. In this installation, the center nailing surface for the top header is the end of a collar tie.

Once the roof has been opened up, the skylight unit can be lifted into place, below. After it is checked for square, it can be fastened to the roof. A helper comes in handy for this part of the job.

and toenail it with 16d nails. Check to make sure you've got a square opening that allows about ¼ in. of clearance on all sides of the skylight unit. Then face-nail through the header into the end grain of the cripple rafters. Toenail the second header to the side rafters, and face-nail it to the first header. After this framing is complete, stuff any remaining openings with insulation, and then make the necessary repairs. If the ceiling is drywall, this is a good time to get a first coat of tape on the patch.

If you are contemplating re-shingling your roof, now is a good time to do it. If not, you will need to remove some of the shingles at the top and sides of the opening in order to flash the new skylight. This isn't a big problem, and it has the advantage that most of the shingles at the sides and bottom of the skylight will be cut and left in position, eliminating the need to mark and trim each shingle individually.

The roof opening should be cut from the outside. Locate the opening by driving nails up through the roof at the four corners of the opening you've just framed up inside. String chalklines and snap the perimeter on the shingles. Then pound the nails back through the roof, and remove them. Now you've got a chalked outline of the rough opening. The shingles should be cut back to about ¾ in. from the edges of the unit, so you need to mark a cut-line ½ in. outside the line you've snapped. Double-check the measurements of both the skylight and the rough opening.

Asphalt shingles are easy to cut with a circular saw and an old carbide-tipped blade. The carbide tips can be dull or chipped, as long as the teeth are widely spaced. Asphalt material quickly gums up ordinary sawblades and makes them useless. Set your blade depth to avoid cutting into the sheathing, then cut the shingles along the outside lines. Remove and discard them. Now mark the rough opening on the sheathing, using as guides the holes left by the corner nails. Cut the roof sheathing out along these lines with a better, sharper blade.

Installing the unit—At this point, the skylight can be set in place. If you're not re-shingling, you'll have to lift some shingles from around the edges of the opening or remove them by popping nails with a flat bar, so that you can mount brackets or install flashing, depending on the design of the skylight. When the unit is centered in the opening, check it for square and be sure it operates correctly. Then attach it securely according to the manufacturer's instructions or the design specifications.

If you are installing a new roof, the new shingles should be applied up to the course whose lower edge is within 10 in. of the opening. The bottom flashing for the skylight can then be installed.

Most skylight manufacturers offer a flashing package with their units. It's usually more expensive than flashing you can make yourself, but if you have a problem with the skylight, the manufacturer could void your warranty if you haven't used his flashing, even if you've done a good job with your own. If flashing has been provided by the skylight manufacturer, install it

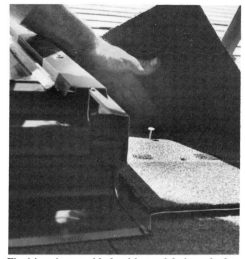

Flashing is provided with prefabricated sky-lights, or can be purchased. Above, the first piece of step flashing is actually an extension of the bottom flashing. It's trimmed so it won't extend below the side shingles, and fastened in place with a nail in its upper outside corner—the proper technique for all step flashing.

With the side shingles trimmed and the top two courses along the skylight's head removed, the top flashing is set in place, right. The top flashing supplied with this unit has a bent return along its upper edge to keep backed-up water from getting under the roofing.

Full shingles are used as the starter course at the top flashing, below right. Like the shingles that will overlay them, they are worked up under those already on the roof, and are nailed high enough so the flashing won't be punctured.

according to the instructions provided. If the instructions merely tell you to nail the flashing to the roof, do so with two nails only, at the upper outside corners of each end.

Some manufacturers supply decent base and head flashing, but rely on strip flashing and mastic along the units' sides. I have little faith in strip flashings, and if that was all the kit included, I would probably make and install my own step flashing, as explained on pp. 76-77.

The inside surfaces of the framed skylight opening can be trimmed with wood or covered with drywall. Depending on the design of the skylight, drywall butting into the bottom of the curb may require a small wood molding to finish it off, or you may decide to flat-tape this gap or rabbet the curb to accept the thicknesses of the finish material.

Your pleasure in looking at the installed skylight for the first time is balanced by your anxiety during the first good rain. No leaks are likely in a proper installation, but you might see wet corners or dripping glass under certain conditions. Condensation is a problem not limited to skylights, and humidity and cold can turn any large expanses of glass into a marvelous condenser. A number of methods exist for controlling humidity in a house, and skylights require no special treatment. If you notice dampness around the skylight, check its source. If it's condensation, forget it.   □

Jim Picton is a carpenter and contractor in Washington Depot, Conn.

Framing for Skylights

The trick is in laying out and building the splayed walls of the lightwell

Skylight frame in place. With the sash removed, the skylight frame can be set on the roof to check the rough opening. The author makes the opening in the roof slightly longer than necessary, then furs out the framing to the correct length. In the photo below, he uses a scrap of ½-in. plywood to determine the thickness of the furring he will nail to the headers at the bottom of the well.

by Doug Hopper

Winter days in the Pacific Northwest can be short and gray, which helps explain why many people rely on skylights to make the most of what little natural light there is. When I started building 20 years ago, "skylight" wasn't even in my construction vocabulary. But skylights are now standard in the houses I build and are one of the improvements homeowners ask for most frequently in remodeling projects.

Framing for a skylight isn't difficult, but it's more than just cutting a hole in the roof and dropping in a factory-made unit. Installation is a lot simpler in new construction because roof and ceiling framing are open and visible, and the skylight can be included in framing plans right from the start. In a retrofit, you'll have to grapple with obstructions like heating ducts, plumbing, wiring

and insulation, and none of these obstacles may be apparent when you start. Whether you're installing a skylight in a new house or an old one, though, framing is essentially the same. There are two basic framing questions you'll have to settle: how to create the openings in the roof and in the ceiling, and how to build the light shaft, or lightwell, that connects the two.

The size of the skylight has a lot to do with how complicated the framing will be. Some skylights fit between rafters or trusses on standard 24-in. o. c. spacing. If you choose one of these units, no rafters or ceiling joists must be cut, and framing is fairly simple. Large skylights require more work and a slightly different framing technique to make the openings and to build the lightwell. I'll explain how I handle both situations.

Planning the skylight well—What makes framing for a skylight unique is the lightwell, which connects openings in the roof and the ceiling. Of course, if there is no ceiling, such as in an open-beam roof, there is no well, and the only required framing will be at the roof. I frame my skylight wells with 2x material so that they can be insulated like an exterior wall (for tips on making well walls from plywood, see sidebar p. 88).

The size of the roof opening is determined by the rough opening of the skylight, but the size of the ceiling opening, and consequently the design of the well, is variable. A simple approach, and the one I use when the well will not be very deep (less than 2 ft. high at the high end), is to bring all four sides of the well down 90° to the roof slope. More often, though, I run the high end of the well 90° to the roof slope, and the lower end of the well plumb. The ceiling opening then becomes longer than the roof opening but the same width as the skylight. The splayed shape allows for greater dispersion of light inside.

This approach works well unless the depth of the well is more than about 4 ft. at the high end. With very deep wells, the ceiling opening can get too long if the high end is run 90° to the roof. An obvious solution is to reduce the angle so that the opening is somewhat shorter or to bevel the high-end wall. A beveled wall starts perpendicular to the roof slope and then breaks about halfway down to continue plumb, or perpendicular, to the ceiling. But before I build that kind of well, I first try to adjust the angles of the well because I think the finished well looks nicer when the walls are each in a single plane.

Another option is to make the ceiling opening greater in width and length than the roof opening. This distributes the light more than the simple splayed-end well, but it's more complicated to build because of the compound angles involved. In addition, it usually means more structural work at the ceiling because you probably will have to cut more joists to make the opening. For those reasons, I don't use this approach very often. Yet it's worth considering, especially when using a small skylight because the distribution of light is increased greatly.

Locating the well inside—In new construction, I often start my skylight framing on the floor below the well, not on the roof. After all of the exterior and interior walls have been laid out on the floor deck, I lay out approximate dimensions for the well on the deck. It may sound backward, but starting the process on the floor lets me see how the skylight well will fit with other openings in the walls, the roof and the ceiling. Sometimes I know how big I want the well to be at the ceiling, but I'm not sure how big a skylight to order. If that's the case, I can use a simple formula to make the calculation (see sidebar this page).

Laying out the well opening on the floor also gives me a chance to make minor adjustments in ceiling and roof framing before it goes up. For instance, a common location for a skylight is centered over the kitchen sink and window. In the case of a single-bay skylight, I may be able to shift roof and ceiling framing a few inches in either direction so that the framing won't encroach on

Sizing a skylight from below

Because I usually build skylight wells with a splayed wall at the high end, the opening in the ceiling will be larger than the rough opening at the roof. What if you know how big the well should be at the ceiling, but you're not sure what size skylight to order? A simple formula can help you solve this layout and framing problem.

Let's say your goal is to frame a 5-ft. long well opening in the middle of a 10-ft. wide ceiling, leaving 2 ft. 6 in. on both sides of the opening (drawing below). Your object is to figure out the rough opening on the roof, assuming the high end of the well is perpendicular to the roof plane, and the low end of the well is plumb. The rough opening will tell you what size skylight to order. The same formula could be used to work the other way, that is, starting with the rough opening and figuring out the size of the well opening in the ceiling.

Here's the formula for figuring out the skylight problem:

$(B \div 12)$ x slope factor = A

A1 in the drawing is the horizontal distance from the outside edge of the rafter to the exterior wall ($10\frac{1}{2}$ in. in this example), the width of the 2x6 stud in the exterior wall ($5\frac{1}{2}$ in.) plus our desired well opening (60 in.) plus the setback from the outside wall (30 in.) for a total of 106 in. B2 in the drawing is 46 in. We also need to know the slope factor, which is the hypotenuse of the triangle formed by the roof. This 9-in-12 roof has a slope factor of 15. (You can get this number right off your framing square in the line named "length of common rafters per foot run.")

It's not hard to plug these values into the formula and get the answers. To determine A2, you use the formula as shown: $(46 \div 12)$ x 15 = $57\frac{1}{2}$ in. To determine B1, you have to invert the formula because the known quantity, A1, is the hypotenuse of the larger right triangle, not the long leg as in the smaller triangle. So, (A÷slope) x 12 = B or $(106 \div 15) = 84\frac{13}{16}$ in. The difference between B1 and A2, $27\frac{5}{16}$ in., is the length of the rough opening. The width of the skylight can vary and doesn't really affect this layout.

With these numbers in hand, I can order the skylight that comes closest to my design goals. I'd wait until I had the unit on the site before I framed the openings and built the well. —D. H.

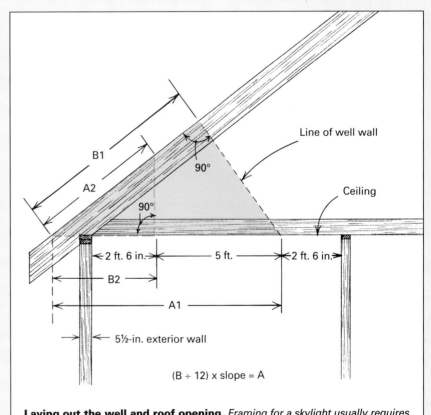

Laying out the well and roof opening. *Framing for a skylight usually requires two different size openings—one in the roof and one in the ceiling. You may start by knowing the size of one of them but not the other. If that's the case, a simple formula can help you calculate the size of the second opening and complete the layout for framing.*

Drawings: Vince Babak

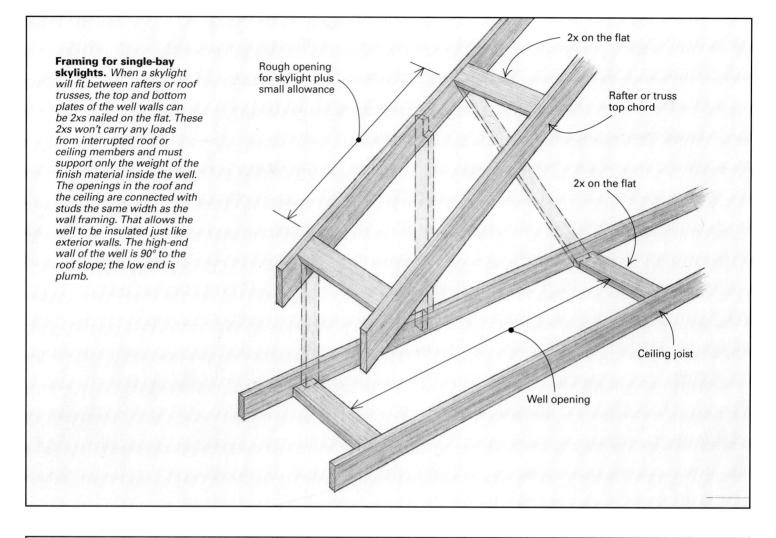

Framing for single-bay skylights. *When a skylight will fit between rafters or roof trusses, the top and bottom plates of the well walls can be 2xs nailed on the flat. These 2xs won't carry any loads from interrupted roof or ceiling members and must support only the weight of the finish material inside the well. The openings in the roof and the ceiling are connected with studs the same width as the wall framing. That allows the well to be insulated just like exterior walls. The high-end wall of the well is 90° to the roof slope; the low end is plumb.*

Rough opening for skylight plus small allowance

2x on the flat

Rafter or truss top chord

2x on the flat

Ceiling joist

Well opening

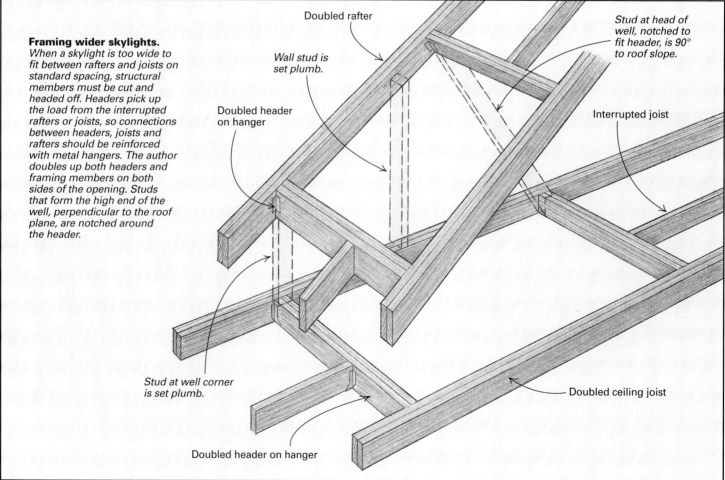

Framing wider skylights.
When a skylight is too wide to fit between rafters and joists on standard spacing, structural members must be cut and headed off. Headers pick up the load from the interrupted rafters or joists, so connections between headers, joists and rafters should be reinforced with metal hangers. The author doubles up both headers and framing members on both sides of the opening. Studs that form the high end of the well, perpendicular to the roof plane, are notched around the header.

Doubled rafter

Stud at head of well, notched to fit header, is 90° to roof slope.

Wall stud is set plumb.

Interrupted joist

Doubled header on hanger

Stud at well corner is set plumb.

Doubled header on hanger

Doubled ceiling joist

the skylight well. This method is a lot faster than framing everything and then going back later to frame in the roof opening—especially when the roof opening wouldn't require headers if located properly. A potential drawback is when a shift in rafter or joist spacing means there will be too great a span for roof sheathing or material I'm going to use on the ceiling. If that's the case, I just drop in an extra rafter or joist to keep spans within allowable limits.

In a retrofit, it may be easier to start the layout at the ceiling. This is especially true when you're trying to avoid an obstruction inside, like a wall, and the location of the skylight on the roof isn't critical. But be careful to avoid vents, roof valleys and other obstructions.

Framing on the roof—Once I know where the well is going, I plumb up from the floor to the ceiling joists and mark one end of the well wall. Those marks can then be transferred to the roof plane, and the roof opening marked on the rafters. I usually add a little bit to the length of the rough opening, 1 in. to 1½ in., and I'll tell you why a little later. The width of the opening should be exactly what the skylight manufacturer specifies. I'll complete the ceiling and well-wall framing from the inside a little later, but once I've marked the rough opening for the skylight on the rafters, I'm ready to frame the skylight opening at the roofline. This is the first step, whether it's new construction or a retrofit.

How I frame the opening depends on whether the skylight will fit between framing members on standard spacing. If the size of the skylight won't require that any rafters be cut, I use single framing members laid flat (parallel to the roof slope) between adjacent rafters to serve as the top plates for the end well walls that I'll build shortly (top drawing, facing page). If it's necessary to cut structural members and head off the framing (top left photo, this page), the standard practice is to double the adjacent framing members (rafters and ceiling joists) that are not cut, and double and set on edge the cross framing, or headers (bottom drawing, facing page).

You should use metal hangers on all connections because the headers are now carrying the load of the interrupted joists or rafters (top right photo, this page). If you have to head off an interrupted rafter, the opening may be too large for the skylight. If so, a single rafter between the headers will narrow the opening to the right size (bottom photo, this page). As a general rule, if I must head off more than one joist or rafter, I'll have the design checked by an engineer.

Finally, the roof sheathing goes on right across the opening. I don't cut the sheathing until I'm ready to install the skylight. That means the job site will be safer because there's one less hole on the roof to fall through, and the house will be less prone to weather damage.

With the rough opening framed in the roof, I can plumb down to the ceiling joists at the low end of the well and mark the location for the wall there (drawing this page). The wall at the high end of the well is perpendicular to the roof, and I use a framing square and a straightedge from the roof to locate the inside of the well wall on

Make way for a skylight. If a skylight can't fit between framing members, a rafter will have to be cut back. Establish the cutline by holding a framing square to the inside of the adjacent rafter and marking the rafter to be cut.

Metal hangers for a strong connection. Because the header is supporting part of the roof load, the author uses a double 2x on edge and connects the header to adjacent rafters with metal hangers.

Establishing the well width. Using a short section of a 2x4, the author wedges a rafter in place and toenails it to the header. He'll add a metal hanger later to help carry the roof loads.

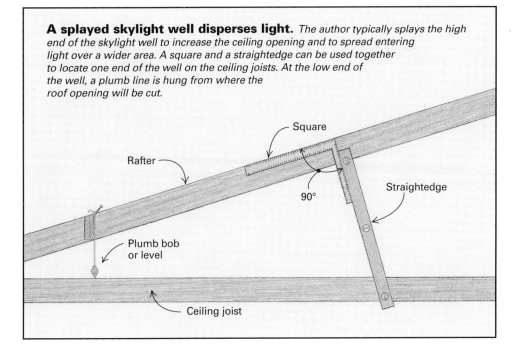

A splayed skylight well disperses light. *The author typically splays the high end of the skylight well to increase the ceiling opening and to spread entering light over a wider area. A square and a straightedge can be used together to locate one end of the well on the ceiling joists. At the low end of the well, a plumb line is hung from where the roof opening will be cut.*

Square

Rafter

Straightedge

90°

Plumb bob or level

Ceiling joist

the ceiling joists. The ceiling opening can then be framed in the same way the roof opening was.

Framing the well—I install the four corner studs of the well after the plates or headers have been nailed between the rafters and the joists. In my part of the country, energy codes require that skylight wells be insulated, so I use the same size framing members for the well as I do for the wall studs (usually 2x6s). For single-bay skylights, with top and bottom plates of the well walls made from 2xs on the flat, the easiest way to measure for the well corner pieces is to cut the top end of the studs at 90° or at the slope of the roof, depending on which end of the well I'm working on. I make the studs a couple of inches longer than their finished lengths. With locations for bottom plates marked on the ceiling joists, but the plates not yet installed, I hold the corner pieces in place and mark them at the bottom

edge of the ceiling joist to get an exact measurement for length. Before cutting, I deduct 1½ in. for the thickness of the bottom plates.

Using the same same technique for determining length and angle, I fill in the side and end framing, spacing the studs just like those in the wall framing (for 2x6s, that would be 24 in. o. c.). On the well's sidewalls, I usually have only a single rafter or joist at top and bottom to which I nail the studs, so I notch the studs over the rafters and the joists to get better nailing.

Framing the well is a little different when structural members, either rafters or ceiling joists, are cut to accommodate the skylight. In that instance, the 2x headers will be on edge. At the high end of the well, the tops of the studs are cut square. The bottom of the studs are notched around the header so that the wall of the well will plane smoothly into the ceiling. The wall at the lower end is framed just the opposite way,

with square cuts at the bottom and angle cuts at the top.

After the well has been framed, I drive nails up through the roof at the corners from the inside, mark the dimensions of the opening on the roof and then cut the opening.

Making final adjustments in the well—Here's why I usually allow a little extra room in the length of the well: The wall on the low end of the well is plumb and therefore intersects the roof plane at an angle. That makes it a little tricky to predict where the line of the wall will emerge through the roof sheathing. It's easier to make the finish opening a little smaller after it's framed than it is to enlarge it once all the framing is in place. So I shim the lower end of the wall out after I've put the skylight on the roof and can measure exactly how much space I have to take up.

I go through this trouble so that the finish material I use on the well lines up with the appropriate spot on the skylight frame. Some skylights have grooves cut in the bottom edges of the frame. These grooves accept the finish material on the sides of the well and make a clean finish. There isn't much room for error in the well framing. After the roof opening has been cut, I remove the sash from the skylight, put the frame on the roof and check the framing (top photo, p. 84). With a fixed unit, this step may require two people: one outside and one inside. If I'm using ½-in. drywall on the inside of the well, for instance, I'll use a scrap of ½-in. plywood as a gauge to align the well framing with the skylight frame (bottom photo, p. 84). At first glance this may not seem necessary, but because of the angle at the bottom of the well, it can be difficult to align the wall solely with a tape or by eye.

Framing in a truss roof—Framing in a skylight on a truss roof shouldn't be a problem as long as the skylight fits between the typical 24-in. o. c. spacing of the trusses. Because you can't modify trusses in the field without consulting an engineer, a skylight that's too big for this opening can be a problem.

I first would use two or more smaller units that would fit within standard framing. The skylights can be ganged together, either side by side or top and bottom, with manufactured flashing designed to span typical framing-member widths. If you place the units side by side, you will build two identical wells separated by a truss.

In new construction, another option is to plan the truss design to allow for a larger opening. This typically means leaving out one truss to create a nominal 4-ft. opening. Be sure to notify the truss manufacturer so that appropriate adjustments can be made to the design (for more on roof trusses, see *FHB* #89, pp. 40-45)

If I'm leaving out a long-span truss for a skylight (and the plan has been okayed by an engineer or the architect), I'll fill the area between full trusses with what I call ladder framing. This consists of 2xs nailed on edge with hangers at each end laid out to align with my sheathing courses. ☐

Doug Hopper is a builder in Tacoma, Wash. Photos by Scott Gibson.

Making a skylight well with plywood

by Larry Haun

Sloped lightwells are open to many variations. They can be sloped on four sides, sloped on one or two sides only, or even sloped more on one side than on another. I've found the easiest way to build the well, especially when it is irregular, is to use sheets of ¾-in. plywood (drawing below) rather than make a 2x frame.

I frame the rough opening in the ceiling ¾ in. larger all the way around. Then, on the underside of the rafter opening, I nail a 2x back from the edge of the end headers and side rafters by at least ¾ in. When the shaft angle is steep, I nail the 2x farther back from the edge so that the plywood sheets will line up with the edge of the roof opening. The steeper the angle of the well, the more setback will be required in the 2x nailer.

Next I cut sheets of plywood that fit from joists to rafters and nail them in

place. This is a good place to use scraps. If the shaft is long, more than 3 ft. or 4 ft., I nail a 2x4 band around the outside of the well to stiffen the plywood structure. Drywall can be nailed or screwed directly to the plywood on the inside, and insulation stapled to the back.

An easy alternative to using 2x nailers is to cut a simple rabbet in the bottom of the rafters and the headers before they are installed. Plywood used for the walls of the well fit into the rabbet. I cut the rabbet with a circular saw and make it at least ¾ in. deep by about 1 in. wide. Make sure the wood is supported and held firmly in place. Once the rabbeted members are in place, the edges of the ¾-in. sheets can be nailed snugly into place.

—Larry Haun is a carpenter in Los Angeles, Calif., and a frequent contributor to **Fine Homebuilding.**

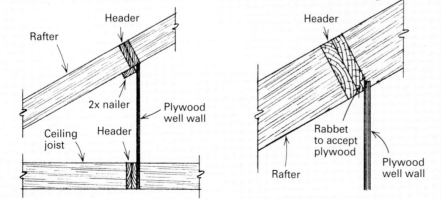

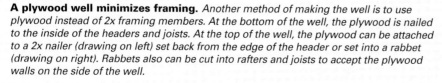

A plywood well minimizes framing. *Another method of making the well is to use plywood instead of 2x framing members. At the bottom of the well, the plywood is nailed to the inside of the headers and joists. At the top of the well, the plywood can be attached to a 2x nailer (drawing on left) set back from the edge of the header or set into a rabbet (drawing on right). Rabbets also can be cut into rafters and joists to accept the plywood walls on the side of the well.*

A Site-Built Ridge Skylight

How an architect built an acrylic skylight for his remodeled kitchen

by David K. Gately

Large residential skylights have always posed problems for architects, contractors and owner/builders. A big custom-manufactured skylight, with tempered or wire glass, extruded-aluminum frame and all the necessary gaskets can be costly. It's also hard to install without a crane and is visually cumbersome—if not downright ugly.

But I think the benefits of a generous skylight far outweigh the disadvantages. It brings the sky, rain, clouds and trees close enough to appreciate. Skylights also serve to reduce the effects of window glare by balancing the natural light in all parts of a room.

I've had some experience fabricating acrylic covers for architectural models, and the thought occurred to me that I could apply the same material and technology to the problem of assembling a site-built skylight. When it came time to design and build my own kitchen addition, I de-

cided to try out a skylight design I've had in the back of my mind for some time. As shown in the drawing on the next page, the design is based on modular panels that interlock, with the flashing edge at the end of one panel nesting in a U-shaped gutter in the adjacent panel.

My case was special in that our house in Mill Valley, California, is well above the road—75 vertical ft. and 135 steps. I needed a skylight material that was light and easy to handle and did not require special technology to fabricate or install. No cranes would come a-calling.

The space that is skylit is a country kitchen 28 ft. long by 15 ft. wide (photo, p. 93). The ceiling is 1x6 cedar decking supported by 2x6 cedar trusses with a roof pitch of 6 in 12. The ridge skylight is 22 ft. long, spanning six 3-ft. 8-in. wide bays (photo below). I wanted to create the most unobtrusive skylight possible and one that also would solve all the prob-

lems of a field-built skylight—namely differential movement of dissimilar materials such as wood, metal and plastic.

Thermal expansion—Movement caused by temperature change is the first consideration in designing an acrylic skylight. The coefficient of expansion of acrylic is .000034 in. per degree of temperature change. This means that in a climate with a yearly temperature range of 20° F to 90° F, the skylight must be designed to accommodate a minimum change in the dimension of each element of 70° F x .000034 in. of expansion x the dimension of the skylight panels. Certain parts of a skylight will heat up well beyond the air temperature. It is wise, therefore, to design for a temperature differential of 100° F in a moderate climate such as ours.

My panels are 44 in. long and 37 in. wide. This means that I must accommodate a length-

The six interlocking segments that compose Gateley's skylight extend 22 ft. along the ridge of the roof. Sunshade cords dangle over acrylic support bars just below the ridge of the skylight. In the foreground you can see the acrylic cap piece and gusset that reinforce the panels.

wise dimensional change of .15 in. (44 in. x 100° F x .000034 in.) and .13 in. widthwise in each panel. To be on the safe side, I designed this skylight for about ¼-in. expansion of each panel in each direction.

Hold-downs—Each panel moves independently with relation to its neighboring panel. The tricky part is attaching the panels to the curb so that they are secured to the roof, yet allowed to move.

My strategy was to use some L-bolt hold-downs that I fabricated in the shop out of ⅜-in. and ¼-in. brass threaded rod (section drawing facing page). The ⅜-in. leg of the bolt extends about 2½ in. into the skylight curb, and the ¼-in. leg fits through the overlapping edges of two skylight panels at the gutters. My skylight has been installed now for a year and a half and has been subjected to strong winds, rain and the full range of temperature change. The hold-downs work well, but I think they are overkill except for skylights that are exposed to high winds. Also, they are tedious to install.

The next time I build one of these skylights, I'll use the L-bolts only at the corners. To link the field panels, I'll use hold-downs that are made of three built-up layers of ¼-in. acrylic. I used some at the bottom of the gutters where they pass over the curb (drawing facing page) and they work well. They are attached to the curb with a bolt made out of ¼-in. brass threaded rod. Above the acrylic hold-down, a ¼-in. brass bolt through the overlapping gutter and flashing edges loosely ties together the adjacent panels. The main point here is that each skylight segment, which is composed of two panels rigidly joined at the peak, must be able to move freely in all directions about ¼ in.

The house is in a relatively protected location and the skylight is not subject to strong winds. A hilltop or ridge location would require additional thought for counteracting 100 mph winds and the resulting uplift. The two hold-down methods described here would suffice for all but the most exposed locations.

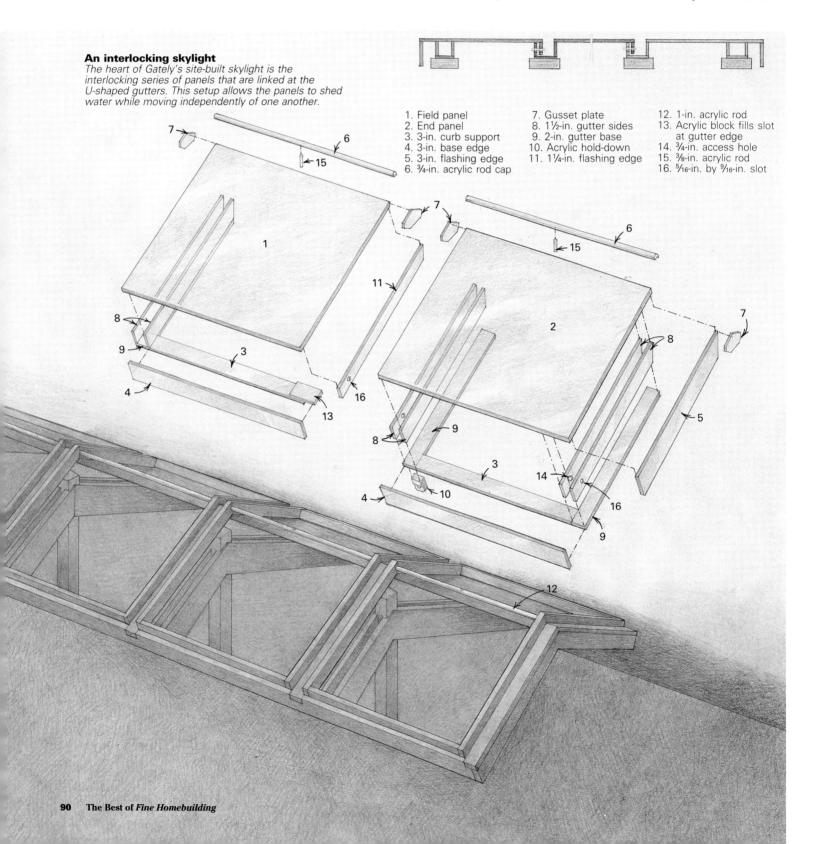

An interlocking skylight
The heart of Gately's site-built skylight is the interlocking series of panels that are linked at the U-shaped gutters. This setup allows the panels to shed water while moving independently of one another.

1. Field panel
2. End panel
3. 3-in. curb support
4. 3-in. base edge
5. 3-in. flashing edge
6. ¾-in. acrylic rod cap
7. Gusset plate
8. 1½-in. gutter sides
9. 2-in. gutter base
10. Acrylic hold-down
11. 1¼-in. flashing edge
12. 1-in. acrylic rod
13. Acrylic block fills slot at gutter edge
14. ¾-in. access hole
15. ⅜-in. acrylic rod
16. 5/16-in. by 9/16-in. slot

Shop drawings—Whether you farm out the acrylic cutting or do it yourself, a descriptive drawing of each component, listing the precise number of pieces required, is essential. A shop drawing need not be to scale and can be done freehand, but it must be precisely dimensioned and clearly marked as to hole locations and bevels. It helps to draw each component on a separate sheet of paper. List the number of pieces to make, assign each one a letter designation, and show a small master-plan view of the whole skylight on each sheet, showing with a darkened line where each piece is located.

Components—The plastic I used was ¼-in. thick cast acrylic with a 10% grey tint. The brand is "Acrylite," which is similar to the brand "Plexiglas." I chose, in the interest of time and because I did not have an adequate table-saw setup, to have the components cut by a local plastics supplier (City Plastics in San Francisco). Only the side pieces terminating at the peak, where they are joined with an opposite panel, need to be precisely cut. Otherwise, the materials are pretty forgiving, and any small discrepancies can be dealt with during installation.

If I had had a good table saw and adequate shop space, I would not have hesitated to cut the components myself, thus saving about $600 for about a day's work. The major tools required are a table saw with an 8-ft. outfeed table, a plastic-cutting carbide planer blade with as many teeth as possible and, for beveling the panels that meet at the peak, a router fitted with a custom bit that matches the plumb cut for the roof pitch (this will cost about $30). Straight, parallel cuts are essential; absolutely smooth cuts are not, although

that would be ideal. The protective paper should be left on during all cutting and beveling. Before you take router to acrylic, be sure to make lots of practice cuts.

Drilling the attachment holes—The holes through which the brass bolts pass should be slotted, as it is here that all the expansion of the material is accommodated. For a slot ⁵⁄₁₆ in. wide by ⁹⁄₁₆ in. long, drill two ⅛-in. starter holes with their centers ¼ in. apart. Then drill each hole partway through with a ⁵⁄₁₆-in. bit, alternating each hole as you go to keep the two holes centered. Keep the protective paper on the plastic as you drill, and clamp the workpiece to a waste piece of thick plastic. Practice drilling a few holes first to get the feel of drilling this material. Acrylic has a tendency to shatter as the bit penetrates the opposite side unless you reduce drill pressure as you finish. Incidently, there are special drill bits designed for working with acrylic. Each has a pointed tip that reduces the likelihood of shattering the material. Plastics suppliers sell them.

Panel assembly—Drilling complete, strip the protective paper from one set of panel components. The only surface left protected is the upper side of the skylight panel. Keep the paper on this surface until the panels are on the roof as this will allow you to slide the panel around the fabrication table and up the ladder.

I assembled the 1¼-in. flashing edges using large spring clamps, and used bar clamps for the 3-in. flashing edges. Once a piece was clamped in position, I applied a solvent called Weldon #4 (IPS, P. O. Box 471, Gardena, Calif., 90247) with a hypoapplicator (a polyethylene bottle with a syringe tip) along the length

of the joint. It's easy to see the spread of the solvent as it moves through the joint via capillary action. Once you've flooded the joint, release each spring clamp briefly to allow the liquid to penetrate the joint fully at that point. Less clamping pressure will result in some excess dissolved plastic protruding from the joint but will produce a stronger joint. Too much pressure will result in a cleaner but slightly weaker joint. This is not critical, but a clear, bubble-free joint is desirable. The process takes about ten minutes per part: three minutes to set up and clamp, one minute to apply Weldon #4 and about four to six minutes for setting and minor adjustments. One or both hands act as additional clamps at this time as you are constantly examining the joint for a clean, transparent set. If you miss a spot, you can always apply more Weldon #4 to achieve a fully welded joint.

Use care with the Weldon #4 liquid, but don't worry if it gets on plastic where you don't want it. Just don't touch spills with anything because that can cause scars. Spills dry rapidly and leave a thin white residue that cleans up with Meguire's Mirror Glaze (Meguire's Inc., 1 Newport Pl., Newport Beach, Calif. 92660).

Assemble the gutters and set them aside. Alignment is most critical where they will be joined at the peak. If they don't quite line up, any overage can be sanded off; underage can be filled during installation with a thick glue called Weldon #16 cement (more on this product later).

Now assemble each field panel, beginning with the 1¼-in. wide flashing edge. Next glue on the 3-in. wide base edge, followed by the gutter assembly. The last piece to be glued in place is the 3-in. wide curb support strip

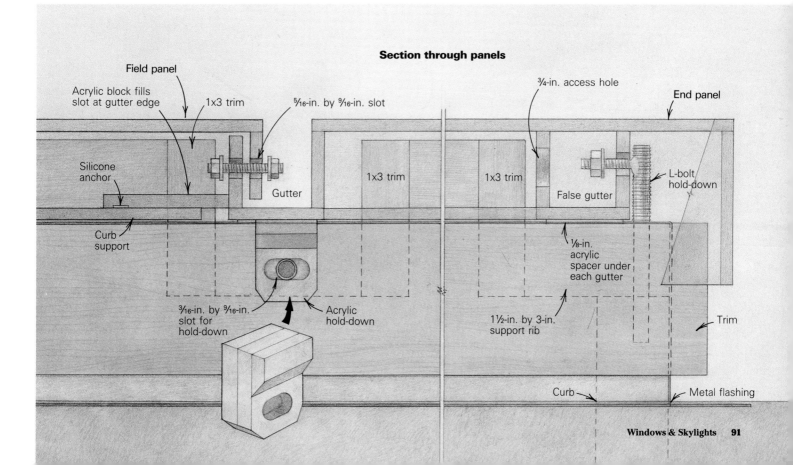

Section through panels

Field panel
Acrylic block fills slot at gutter edge
1x3 trim
⁵⁄₁₆-in. by ⁹⁄₁₆-in. slot
¾-in. access hole
End panel
Silicone anchor
Gutter
1x3 trim
1x3 trim
False gutter
L-bolt hold-down
Curb support
⅛-in. acrylic spacer under each gutter
³⁄₁₆-in. by ⁹⁄₁₆-in. slot for hold-down
Acrylic hold-down
1½-in. by 3-in. support rib
Trim
Curb
Metal flashing

(drawings, pp. 90 and 91). It abuts the gutter at one end of the panel, but stops 1 in. shy of the flashing edge at the opposite end of the panel. The resulting notch allows the panel to be fitted over the interlocking edge of the gutter. To position the curb support piece, scribe a fine line on the base edge using a razor knife guided by a straightedge.

The field panels each have a gutter along one side and a flashing edge on the opposite side. The end panels, however, are a little different. They have a 3-in. wide flashing edge that runs from the base edge to the peak. Also, they have a "false gutter" that rests atop the curb assembly at either end of the skylight well (drawing previous page). The false gutter acts as a box beam to strengthen the panel, and as a spacer to keep everything in the right plane. The right end panel has a gutter assembly on its left side. The left end panel has a flashing edge on its right side.

When all the panels are assembled, hand-sand any parts of the gutter assemblies that project past the peak. I sanded all the edges that meet at the peak, including the bevel on the panels, to make these pieces even. A sanding block with #80-grit paper contact-cemented to a thick piece of scrap acrylic works well.

Installation—With the paper still on the upper face of the plastic, move the panels onto the roof. I just pushed them up the ladder, paper side down. Although they are not fragile, one must take care not to torque them, which may loosen a joint. If this does happen, just apply some more Weldon #4 to the separated joint, and hold it steady until it sets

up again. Each panel weighs perhaps 20 lb. and is slightly cumbersome, but it is still just a one-person job to move one.

At this point place the ⅛-in. thick acrylic spacer/bearing strip on the tops of the support ribs. I used small finish nails along the sides of the ribs to retain them, leaving just enough of the nail head exposed to catch the edge of the acrylic strips. Now is also the time to affix weatherstripping to the metal flashing atop the curb (drawing below). I used the self-sticking neoprene foam type, ⅜ in. wide by ¼ in. thick.

Attachment—The panels are anchored with the L-bolt hold-downs at each corner. At the end panels, the ¼-in. dia. leg of the hold-down extends into the false gutter, where it's reached from inside the skylight well through a ¾-in. hole drilled through the inboard edge of the false gutter. The brass nut and washer are installed through this access hole. I put a little contact cement on the washer, which temporarily stuck it to the socket during installation.

I began the installation with the right-side end panels. With the two panels sitting in the correct position, their peak edges aligned, I clamped them together with several elastic cords linked together, stretched over the peak and hooked to the base edge of each panel. Panels aligned, I marked the curb for placement of the L-bolts at both leading and trailing edges of the panels, leaving enough space between panel and bolts for longitudinal expansion of about ¼ in. Transverse expansion is accommodated by the slotted holes. Then I removed the panels, drilled my holes and installed the L-bolts.

The horizontal leg of the L-bolt must be short enough to allow the panel to be threaded onto it, but long enough to accept a washer and nut (about ⅝ in.). The panels are still unjoined at this point so there is a lot of play to accomplish this maneuver. The L-bolts can also be turned a little to help positioning. To align some of the bolts, I reached through the access holes with a hooked wire.

Once I had the first two end panels in place, the rest of the panels were quite easy to position. At this stage it is a simple matter to adjust the panels for alignment by finger-tightening the nuts from the outside while kneeling on the roof. Don't be alarmed if the ridge edges of the panels sag when first placed. They will become straight, rigid and structural when joined with Weldon #4 glue.

Once I had all the panels in place, I removed the protective paper. This is a nice moment of revelation because the light is finally allowed to stream into the room. But the skylight is not watertight yet, and in this case, with night coming on, I ran duct tape down the ridge in case of rain. Rain it did, and the tape kept it out.

Joining the ridge—Working from the outside, I used two trios of elastic cords to hold opposing panels in place at their ridge ends during glue-up. To prevent scratches, I wrapped the hooks on each cord with duct tape.

I used an awl inserted vertically between the panels to adjust the alignment of their edges. If one of the panels is a little low, lifting on the awl brings it back in line. With the ridge and panel ends precisely aligned and under tension from the elastic cords, I applied Weldon

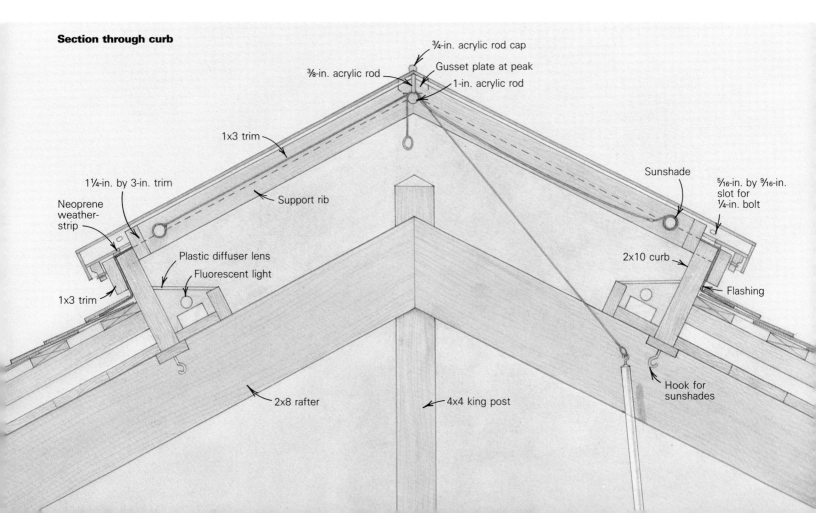

Section through curb

¾-in. acrylic rod cap
⅜-in. acrylic rod
Gusset plate at peak
1-in. acrylic rod
1x3 trim
Sunshade
1¼-in. by 3-in. trim
⁵⁄₁₆-in. by ⁹⁄₁₆-in. slot for ¼-in. bolt
Neoprene weather-strip
Support rib
Plastic diffuser lens
Fluorescent light
2x10 curb
1x3 trim
Flashing
2x8 rafter
4x4 king post
Hook for sunshades

#4 liquid to the ridge joint with the hypoapplicator. I started at the center of each panel and worked outward, flooding the joint with the solvent as far as I could reach. Then I pressed the ridge joint together with my fingers and held it briefly until it took a set. I didn't worry about joining the entire length all at once, and I found that a few inches of the joint at each end wouldn't entirely close. I filled these gaps with Weldon #16 clear cement. This is a thickened, syrupy version of Weldon #4 used for high-strength joints that need some filler. This cement flows fast, so crimp the nozzle to $\frac{1}{16}$ in. or less to reduce flow.

After each succeeding pair of panels was joined with Weldon #4 and the ends were filled with Weldon #16 cement, I applied a small bead of Weldon #16 to the outside and inside of the skylight peak along its entire length. This is really the only part of the gluing process requiring any degree of skill. A little practice reveals the behavior of the cement under vertical and horizontal conditions; it hardens at a certain rate and flows from the tube evenly if applied at an even speed. Uneven flow can be corrected while the cement is still liquid by backing up slightly and stretching out the bead while slightly increasing the flow. An even, $\frac{1}{8}$-in. wide bead is the goal. Weldon #16 will occasionally whiten as it cures. This happens unevenly, so use it sparingly where it shows from below.

To finish the installation I capped the notches that remained in the curb supports (next to the gutter assemblies) with a square of $\frac{1}{4}$-in. acrylic, secured with a dab of silicone caulk (drawing, p. 91). Next time I trimmed out the support ribs with 1x cedar. To clean up the scuffs and dribbles of installation, I polished the main panel surfaces both inside and out with Meguire's #17 cleaner, followed by an application of Meguire's #10 mirror glaze.

The test of summer—When I initially discussed the structural feasibility of my ridge joint with the plastic supplier, we both thought that a plain butt joint would be adequate. In the event that it was not, a ridge cap could be installed to solve any structural problems that might arise. Sure enough, during the hot summer weather, each opposing pair of panels expanded and contracted so actively that the unreinforced ridge joint gradually failed, starting at the ends and working inexorably toward the middle.

To counter the forces involved, I added gussets to the vertical gutter sides at the ridge (drawings, p. 90 and facing page). To further strengthen the ridge and create an absolutely watertight joint, I added a cap made from a $\frac{3}{4}$-in. acrylic rod V-grooved lengthwise so it conforms to the shape of the peak. These solutions have worked well, and should be considered an integral part of the design. They also eliminate the need for the meticulous application of Weldon #16 cement.

An inevitable by-product of a clear skylight is the summer sun's effect on things such as candles, butter and fabrics, namely wilting, melting and fading. I had planned from the start to install some form of shading device, and living with the sun hastened the need.

I settled on a design that complements the lines of the acrylic cap piece. A 1-in. dia. acrylic rod is now located in each skylight bay, 1½ in. below the peak (drawings, p. 90 and facing page). The rod is let into and supported by the wood trim. It flexes slightly under the pressure of drawing the shades, so I installed a ⅜-in. vertical rod at the center of the span. It is let into ½-in. deep holes drilled into the ridge cap and the 1-in. rod, and it's affixed with Weldon #4. The shades are mounted to the sides of the trim pieces (drawing facing page), and their pulls are draped over the 1-in. acrylic rod. To close the shades, I use a 7-ft. pole with a hook at the end and slip the shade pulls over hooks screwed into the ceiling trim.

The cost—I priced a custom metal and glass ridge skylight at $3,500, single-glazed, not including installation which would have been costly—probably another $1,000. My cost for acrylic materials was about $1,000 for 132 sq. ft. of skylight, or $7.50 per square foot. Fabrication took me about 20 hours and installation about 24 hours, which included plenty of head-scratching as I puzzled out the assembly process.

The whole assembly of skylight and shades is now fully operational and I anticipate years of trouble-free service. Since the acrylic panels are so active, especially during the summer, I plan to inspect the edge joints on the panels when I make my annual autumn trip to the roof to clean the gutters. If I find any joints that show signs of opening, I know I can reseal them with a squirt of Weldon #4 from the hypoapplicator. □

David K. Gately is a partner in the architectural firm of Callister Gately Heckmann & Bischoff in Tiburon, California.

Rows of red cedar trusses are illuminated by the ridge skylight in the author's kitchen. To keep the room from overheating on hot days, Gately installed retractable shades near the skylight curbs. They can be opened and closed manually with a pole that has a hook mounted on the end.

Drawings: Gary Williamson

Curbless Skylights

Insulated glass mounted flush with the roof in a low-cost, site-built design

by Rob Thallon

Builders find themselves installing lots of skylights these days. There are some excellent manufactured skylights on the market, the best of which are well insulated and can be opened for ventilation and cleaning. These top-of-the-line units are expensive—$25 per sq. ft. or so. Less costly store-bought skylights are not operable, usually not insulated, and they are often translucent rather than transparent. Some of these budget skylights come with an attached, pre-flashed curb to raise the glazing above the roof, but many require on-site curb construction.

The curbless skylights that I started to build about nine years ago are installed more like large shingles, in contrast to most of the manufactured or site-built skylights that I've seen. As the photo at right shows, eliminating the curb gives the skylight a lower profile so that it looks more continuous with the roof.

In the last six years my partner, David Edrington, and I have installed more than 50 curbless skylights, and they've held up well. We use insulated, tempered-glass panels in standard sizes when possible, and we've simplified construction details to the point that we now feel confident that our skylights are the best fixed skylights available.

Our design can accommodate just about any type of glass or acrylic panel (providing you allow extra room for expansion), and replacing the glass is an easy job. The design doesn't rely on caulks and sealants, which have unpredictable lives, but rather on the behavior of water in contact with metal and glass. With slightly different flashing details, these skylights can be ganged to form continuous bands of roof glass such as those found in greenhouses. And perhaps best of all, our skylights cost about $4 per sq. ft., if you use off-the-shelf sizes of insulated glass.

General suggestions—The glass and side flashing drains in our system are at a slightly lower angle than the roof, so the amount of pitch is important. We've used this detail in 4-in-12 roofs, but I don't recommend going any shallower than this.

The skylight construction details can be adjusted to work with most roofing materials. We try to use those that fit closely, like cedar or composition shingles. Roll roofing or metal

Architect and builder Rob Thallon lives in Eugene, Oregon. Photos by the author.

would also work fine, but with looser-fitting materials like shakes or tiles, the step and side flashing dimensions should be increased.

Organizing materials—Since the size of the glass determines the size of everything else, this is the first part of the skylight to consider. Insulated tempered-glass door blanks come in a standard 76-in. length and in three standard widths—28 in., 34 in., and 46 in. We usually use the 34-in. wide blank, and the installation I'll describe uses this size.

You can also adapt these instructions for glass of any size. We've had smaller tempered windows custom-built, but they end up costing more than the larger ready-made ones. On the other hand, glass larger than 34 in. by 76 in. usually isn't strong enough to be safe. Consult your glazing supplier about the panel strength in relation to size and snow loads in your area, and check with your building inspector for the minimum glass-thickness requirements for skylights.

It's important to understand that ordinary window glass isn't recommended for skylights because of its relatively low strength and because, if it should break, a large piece of it falling into a living space could be extremely dangerous. The three types of glass recommended for skylights—tempered glass, safety glass and wire glass—have overcome this problem in different ways. When it breaks, tempered glass is supposed to dice into tiny

bits, each no larger than 3/16 in. Safety glass (also called laminated glass) is a sandwich of two layers of ordinary glass held together by a layer of plastic. Wire glass has a network of tiny wires running through it that prevent the glass from breaking into dangerous shards.

Of all the kinds of glass, we prefer the tempered for skylights because it's the most transparent and, under uniform conditions, its strength exceeds that of laminated or wire glass by a factor of about four. The chief disadvantage of tempered glass is the potential for incomplete dicing when it breaks (for more on the dangers of tempered glass, see *FHB* #16, p. 21). We've never had such a problem with any of our skylights, but the possibility has emerged, so codes governing tempered-glass skylights have been stiffened here in Oregon. Building departments require insulated units to have a safety-glass layer on the inside or a screen below tempered glass. The screen has to be at least 12 USA-gauge wire with a mesh no larger than 1 in.

Once you've decided on the type and size of the glass, you can order the flashing. General dimensions for each flashing configuration are noted in the bottom drawings on the facing page; the dimensions in parentheses are for a 34-in. by 76-in. unit. The bends are straightforward and should be easy work for any reputable sheet-metal shop. We usually specify 26-ga. galvanized steel or 16-oz. copper, but prepainted or stainless-steel flashing will work fine. We don't use aluminum because it won't bend to these shapes without fracturing.

At our local shop, the flashing package for a 34-in. by 76-in. unit costs $32.75 for galvanized steel, $81.00 for copper, $42.50 for prepainted and $77.00 for stainless steel.

Roof framing and flashing—The rough opening for the skylight needs to be ½ in. wider (across the roof) and 2½ in. shorter (parallel to the rafters) than the dimensions of the glass. For example, the rough opening for a 34-in. by 76-in. glass panel should be 34½ in. by 73½ in.

Sheathe the roof to the edges of the rough opening (but don't let the sheathing project into the opening), and install the roofing material up to the bottom edge of the opening. Next, fasten 1x4s to each side of the rough opening so their top edges project a couple of inches above the sheathing. These are just temporary fences for the step flashing, but

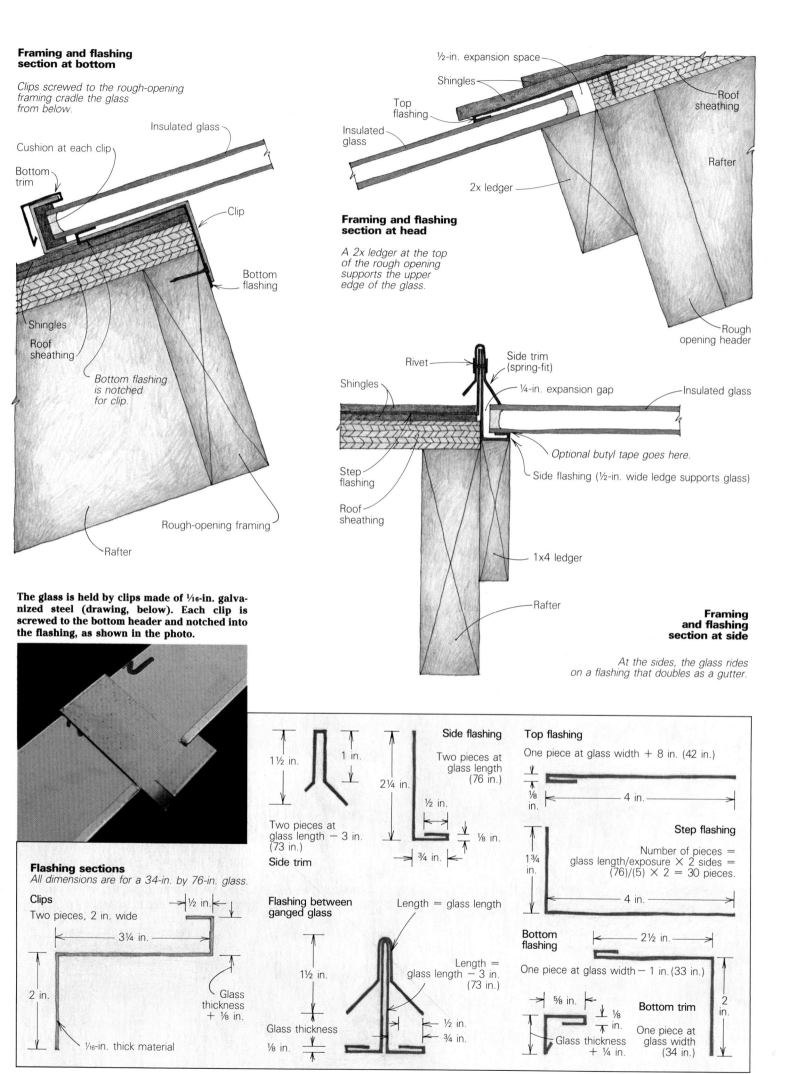

Framing and flashing section at bottom

Clips screwed to the rough-opening framing cradle the glass from below.

Insulated glass

Cushion at each clip

Bottom trim

Clip

Bottom flashing

Shingles

Roof sheathing

Bottom flashing is notched for clip.

Rough-opening framing

Rafter

½-in. expansion space

Shingles

Top flashing

Insulated glass

Roof sheathing

2x ledger

Rafter

Framing and flashing section at head

A 2x ledger at the top of the rough opening supports the upper edge of the glass.

Rough opening header

Rivet

Side trim (spring-fit)

Shingles

¼-in. expansion gap

Insulated glass

Step flashing

Roof sheathing

Optional butyl tape goes here.

Side flashing (½-in. wide ledge supports glass)

1x4 ledger

Rafter

Framing and flashing section at side

At the sides, the glass rides on a flashing that doubles as a gutter.

The glass is held by clips made of ¹⁄₁₆-in. galvanized steel (drawing, below). Each clip is screwed to the bottom header and notched into the flashing, as shown in the photo.

Flashing sections
All dimensions are for a 34-in. by 76-in. glass.

Clips
Two pieces, 2 in. wide

½ in.

3¼ in.

2 in.

¹⁄₁₆-in. thick material

Glass thickness + ⅛ in.

1½ in.

1 in.

Two pieces at glass length − 3 in. (73 in.)

Side trim

2¼ in.

½ in.

¾ in.

⅛ in.

Side flashing

Two pieces at glass length (76 in.)

Flashing between ganged glass

1½ in.

Glass thickness ⅛ in.

Length = glass length

Length = glass length − 3 in. (73 in.)

½ in.

¾ in.

Top flashing

One piece at glass width + 8 in. (42 in.)

⅛ in.

4 in.

Step flashing

Number of pieces = glass length/exposure × 2 sides = (76)/(5) × 2 = 30 pieces.

1¾ in.

4 in.

Bottom flashing

One piece at glass width − 1 in. (33 in.)

2½ in.

2 in.

⅝ in.

⅛ in.

Glass thickness + ¼ in.

Bottom trim

One piece at glass width (34 in.)

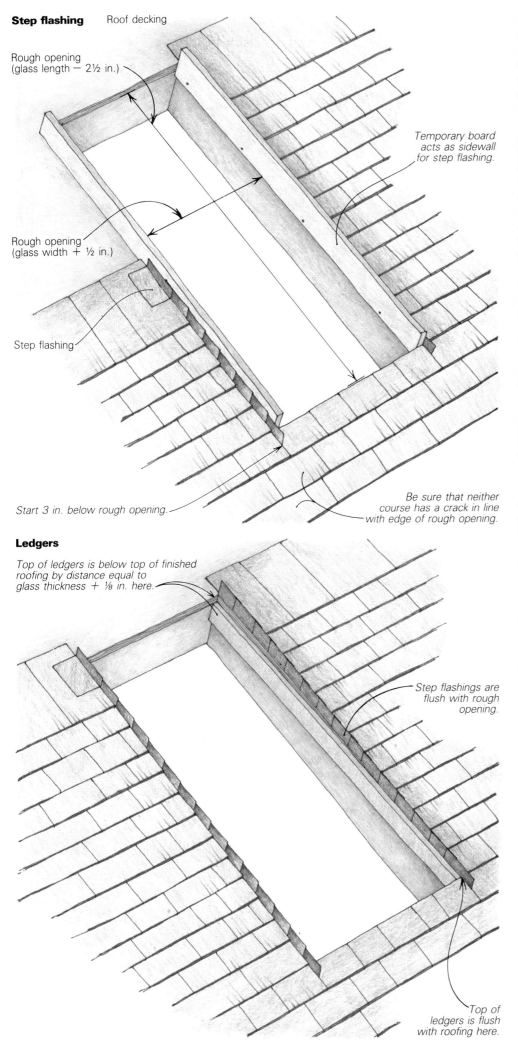

Step flashing

Roof decking

Rough opening
(glass length − 2½ in.)

Rough opening
(glass width + ½ in.)

Temporary board
acts as sidewall
for step flashing.

Step flashing

Start 3 in. below rough opening.

Be sure that neither
course has a crack in line
with edge of rough opening.

Ledgers

Top of ledgers is below top of finished
roofing by distance equal to
glass thickness + ⅛ in. here.

Step flashings are
flush with rough
opening.

Top of
ledgers is flush
with roofing here.

they should be fairly secure just the same, since you'll use them as though they were sidewalls. Continue roofing up the sides of the opening, installing the step flashing tight against the 1x4s (drawing top left). When the step flashing reaches the top of the rough opening, remove the temporary sidewalls.

Next, nail permanent 1x4 ledgers to the sides of the rough opening (drawing, bottom left). These should be cut from dry wood that has a moisture content of less than 15%, to reduce any chance of warpage. At the downhill end of the opening, make these side ledgers flush with the installed roofing. At the uphill end, set the ledgers below the surface of the roofing by a distance equal to the thickness of the glass plus ⅛ in. This is very important. These ledgers can be used as finish trim inside the opening, so you might want to use clear material and rip these pieces so they're flush with the ceiling finish.

At the top of the rough opening, install a 2x ledger with its top edge parallel to the angle of the roof, and set ⅛ in. above the side ledgers already installed. This detail is shown on the facing page, drawing top right.

Now you can install the flashing for the glass. The side flashing butts against the already installed step flashing, with its uphill end at the top of the rough opening and its bottom projecting 3 in. over the new roofing. The side flashing has a J-shaped profile, which creates a small gutter between the top of the side ledger strips and the sides of the glass panel. This gutter acts as a runway for moisture that gets past the counterflashing.

The side flashings don't need to be nailed to the ledgers to hold them in place, but if the roof is especially steep you might want to tack each one down at the very top to hold it steady during assembly. Use tin snips if necessary to trim the top of the side flashing flush with the top edge of the step flashing.

Next comes the only tricky part of the operation. Using tin snips, cut a notch out of the step flashing 3 in. down from the top of the rough opening and flush with the roofing, as shown in the top drawing on the facing page. Make a vertical cut in the side flashing 2¾ in. down from the top of the rough opening and fold the resulting flap onto the adjacent step flashing. There is now a ¼-in. wide tab in the side flashing that should be folded around the step flashing for a mechanical connection.

The tiny gap between the step and side flashings needs to be caulked or soldered against the weather. Soldering, which works on copper and galvanized, is more permanent than caulking, but also more difficult, so we usually seal this tiny crack with a gutterseal caulk made for galvanized gutters.

The bottom flashing is next. Cut out two notches in the bottom flashing wide enough for the panel clips to pass through and about 6 in. in from the bottom corners of the panel. Lay this flashing across the bottom edge of the rough opening and install the clips in their notches by screwing them to the header at the bottom of the rough opening (photo, previous page). Now you can install the glass.

Placing the glass—The glass panel is laid so that its edges rest on the side flashing and its bottom edge is supported by the clips. Neoprene cushions (setting blocks) inside the clips will lessen the chance of the clips starting a crack in the glass. We usually use the pads that protect the glass as it comes from the factory for this purpose.

As you get ready to install the glass, remember that a standard 34-in. by 76-in. double wall unit weighs about 105 lb. Be sure your footing is solid, and have as many hands available as is practical.

Here in Oregon, we haven't had problems with excessive air infiltration, but people in cold climates might want to bed the glass on a perimeter of butyl tape. Use narrow (¼-in.) tape to avoid clogging any of the drainage channels, and don't let the butyl tape come in contact with the edge seals of the glass panel—they aren't compatible and may cause each other to deteriorate.

With the glass in place, you can install the top flashing. (For installation details of the top, side and bottom flashing and counterflashing, see the drawings at the top of p. 95.) The top flashing fits 3 in. over the top of the glass, with its bottom edge resting at the notches in the side flashing. Lay this top flashing in place and tack it to the roof sheathing with a couple of nails near its top edge. Now finish the roofing, and be sure not to put any nails into the part of the glass that's hidden under the flashing.

The last step is installing the counterflashing. First slide the bottom trim piece over the clips so that it covers the bottom edge of the glass and is supported by the clips. This shields the sealant at the panel's bottom edge from the ultraviolet rays of the sun. Next slip the counterflashing pieces over the lip formed by the step flashing and the side flashing. This counterflashing is spring-fit against the glass. You'll have to cut out a small notch at the bottom of the counterflashing where it passes over the bottom trim. Fasten step flashing, side flashing and counterflashing together with three pop rivets per side—about 3 ft. o. c. If you're using galvanized flashing, dab caulk on these rivets to prevent rust.

If you want to make a skylight larger than the size of a standard tempered-glass unit, your best bet is to gang several together. We've done this often in solariums and greenhouses. The details remain the same except between adjacent pieces of glass. The flashing for this condition is a twin piece of side flashing (see p. 95, drawing bottom center). It's treated at the top and the bottom exactly like the side flashing already described. If, as is often the case in greenhouses, you want the glass to come right to the eave and drain directly into the gutter, you can eliminate the side ledgers, and just use the rafters to support the glass and flashings. This works especially well when the thickness of the roof sheathing approximates the thickness of the glass, as it often does. Here the glass can lie directly on the rafters, which can be finished to be an integral part of the installation. □

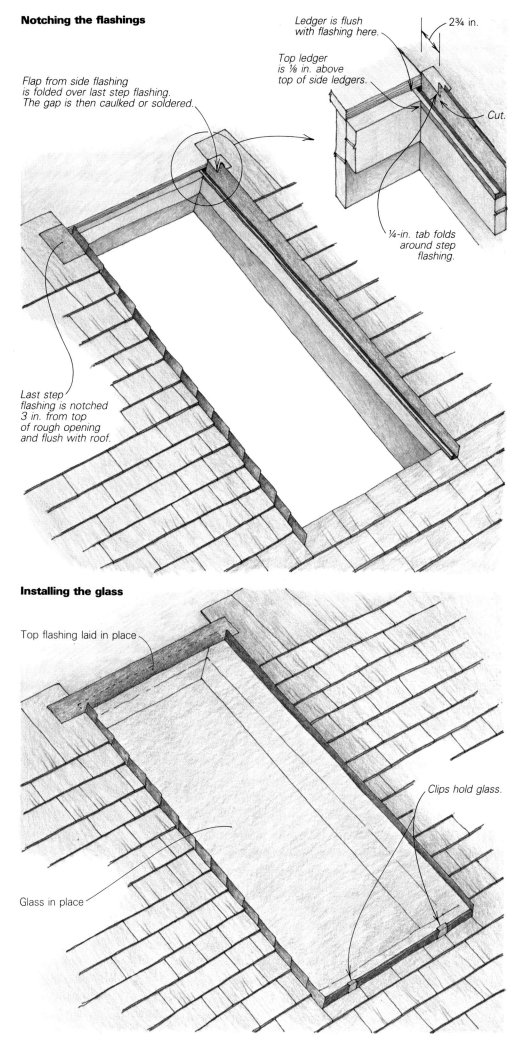

Notching the flashings

Flap from side flashing is folded over last step flashing. The gap is then caulked or soldered.

Last step flashing is notched 3 in. from top of rough opening and flush with roof.

Ledger is flush with flashing here.

Top ledger is ⅛ in. above top of side ledgers.

2¾ in.

Cut.

¼-in. tab folds around step flashing.

Installing the glass

Top flashing laid in place

Glass in place

Clips hold glass.

Skylights in the Eaves

Creating private, light-filled spaces when houses are close together

by Anthony Simmonds

Daylight and privacy. **An alcove illuminated by side windows and a skylight now occupies the portion of wall that used to contain a large window overlooking the neighbor's living room.**

A row of 24-ft. wide houses on 33-ft. wide lots leaves little in the way of sideyards between the buildings. Yet many older houses on these narrow lots still have big windows—often quite grand windows—in their long sidewalls. Winter days in Vancouver tend to be short, damp and dark; we need all the light we can get. Sidewall windows, however, are usually less than satisfactory. They typically provide an unobstructed view of the sidewall of the next-door neighbor's house, a view that may include the corresponding window in his sidewall—in which case the problem of an unattractive vista is aggravated by a lack of privacy.

I get on well with my neighbors, and feel fortunate indeed to be living in a house flanked by families who like each other well enough to abolish two fences and share the one resulting oversize backyard. But I still wanted an alternative to the 5-ft. by 7-ft. window, stained-glass and all, that gaped at the dark brown sidewall of Peter and Susan's house 12 ft. away. Admittedly, the corresponding window in their sidewall did afford a means of mimed communication for the children. I, however, didn't feel communicated with—I felt perused.

The solution I came up with is a refinement of an idea mentioned to me by an architect friend soon after I bought the house. He had suggested taking advantage of the large opening already framed in the wall by pushing out a small "blind" bay that would have narrow windows in its ends rather than its face (photo left). In addition to giving us some privacy, the bay would make a space large enough for a couch, or even an upright piano, without crowding the big table that is the center of our large farmhouse-style kitchen.

When I began the project, it occurred to me that I could take advantage of another feature of the house: its generous roof overhang. On my house this overhang is 2 feet wide, quite enough to accommodate a conventionally framed 2x4 wall with its sheathing and siding, as well as a tall, narrow window with no trim. This would obviate the need to build an independent roof for the bay (drawing facing page). Best of all, the deep eave allowed me to let in a great deal more light simply by

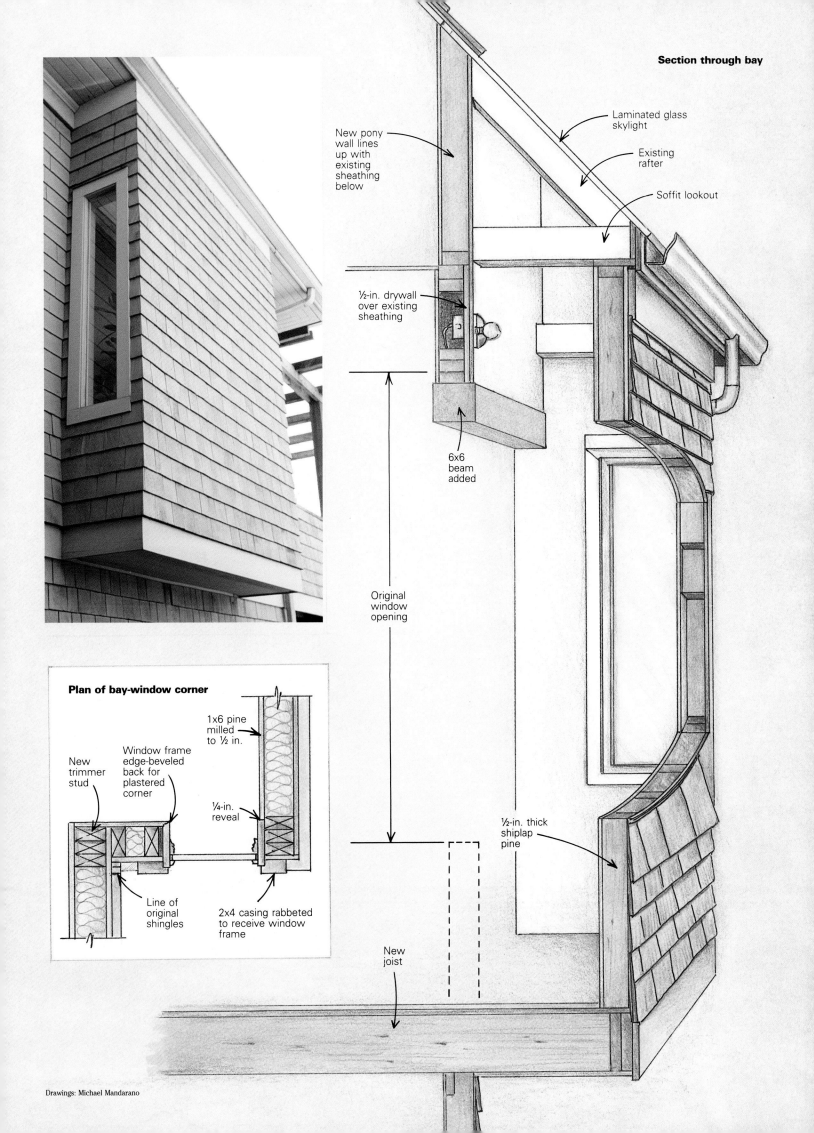

Section through bay

Laminated glass skylight

Existing rafter

Soffit lookout

New pony wall lines up with existing sheathing below

½-in. drywall over existing sheathing

6x6 beam added

Original window opening

½-in. thick shiplap pine

New joist

Plan of bay-window corner

1x6 pine milled to ½ in.

Window frame edge-beveled back for plastered corner

New trimmer stud

¼-in. reveal

Line of original shingles

2x4 casing rabbeted to receive window frame

Drawings: Michael Mandarano

opening the soffit and a section of the roof above and replacing it with glass.

Extending the floor—In order to have a platform on which to work, I began by extending the main-floor joists. The basement ceiling was unfinished, so this presented no problem. In fact, it was made especially easy because the house is balloon-framed, with the studs running full length from foundation to top plate, and the first-floor joists supported by a 1x4 ribbon-board let into the studs. This meant I had no rim joist to cut through, and once the wall framing underneath the window was exposed it was a simple matter to determine which floor joists had to be extended. The joists closest to either end of the bay would have to be furred out to provide solid bearing under the end walls, but I left this until the exact size of the opening could be determined.

Working from the outside of the house, I used a circular saw to cut slots in the wall adjacent to each of the joists that had to be extended. These slots started out at an optimistic 1¾ in. by 7¾ in., but it's surprising how shy a slot like that is—it positively shrinks when you introduce a 2x8 to it. So after a good deal of fiddling and cursing on the first one, I enlarged the others somewhat.

The new 2x8 joists had to extend less than 2 ft. beyond the wall, but in order to provide a comfortable platform to work on while removing the big window, I let them run a foot longer. I used 14-ft. 2x8s, which were long enough to attach to the center girder of the house. The new joists, like their ancestors, were supported on the 1x4 ribbon, but with pianos in mind, I added trimmer studs that extended to the sill plate under each new joist.

Once the window was removed, along with the section of wall underneath it, I could size up the bay's dimensions. At the top, I left the section of wall over the window intact. The economy of using an existing opening in an exterior bearing wall obviously required leaving the header in place, but I also wanted the head of the bay to look quite low from inside the room—not above door height. Achieving this actually meant lowering the top of the opening, which I did by putting in a 6x6 beam right underneath the existing header. It bears on two new trimmer studs either side. The beam echoes another, deeper beam in the center of the kitchen, and fulfills code requirements for beefing up the old framing.

But those trimmer studs had another purpose. Installing them allowed just enough room at each side for 2x4 framing and ½-in. plywood sheathing to tuck in neatly behind the existing line of the shingles (drawing previous page). All I had to do was extend the vertical cut lines on either side of the window up to the soffit, and strip away the shingles in between. The last few inches of fascia board and crown molding, which my saw couldn't reach, yielded to hammer and chisel, jigsaw, pruning saw and general beavering. It would have been easier, of course, to get at this last bit by cutting through the soffit in the same line and coming at it from above. The cut line in the soffit, however, needed to be 3 in. or so to the inside of the bay, thus providing a flange for screwing the soffit boards to the top plate of the new walls.

Enclosing the bay—With the soffit removed, the walls could be framed and the bay enclosed. It was time to contemplate the size of the windows in the end walls. I wanted to maximize the area of glass, yet ensure that the windows would be in keeping with the rest of the house. To do so, I decided that as far as the exterior was concerned, they must have at least a 4-in. wide casing, and that some shingles must show on either side of that. This presented no difficulty at the outside corner, where placing the window frame directly against the inside of the long wall of the bay would allow space for a rabbeted 2x4 casing and a woven shingled corner beyond it. If I made the same allowance at the inside corner, however, I would be left with a rough opening some 12 in. wide. I had originally visualized casements, but I must have been dreaming—even with minimal allowances, the glass would have been less than 6 in. wide.

The interior posed a different problem. The window was jammed against the outer wall of the bay, leaving no space for any casing at all, let alone a wide one. Clearly it was impossible to center the window in both inside and outside walls. A compromise, with space for a narrower casing at the juncture of window and outside wall, would sacrifice too much glass area for the sake of a less-than-ideal detail. I decided to leave the window pushed up against the outer wall and to have no side casings at all, just a lintel to define the top.

With this decision made, I framed the openings and ordered the double-glazed window units. The walls were quickly enclosed, and I now turned my attention to the roof.

Into the roof—I wanted to avoid structural alteration to the roof as much as possible and take advantage of the existing framing to support the skylight. I had used the "curbless skylight" detail devised by Rob Thallon (pp. 94-95) with good results on other jobs and planned to use it atop this bay. His method uses a system of flashings and sheet-metal flanges placed on the rafters to carry the skylight glass (for more on skylights, see "Building Doors, Windows and Skylights," *The Builder's Library*, The Taunton Press, 1989). I like the low profile of the curbless skylights, but one could install the more common curbed variety over a bay like this, though not without sacrificing some of the glazed area.

I knew from previous forays into the eaves that the rafters were in line with the second-floor joists. However, there was no reason to suppose that these would have any relationship to the old window opening because it had been centered in the wall rather than centered according to stud layout. So I considered myself lucky when, on removing the soffit, I found a rafter within 6 in. to 10 in. of

Plants in the eaves

Our stairwell presented a problem similar to one we had encountered downstairs. The well was lit by a single window in a dormer that directly faced a similar dormer in my neighbor's house. So I took out the old window, replaced it with siding and put in a big skylight that made the stairwell wonderfully bright. But in the course of stripping lath and plaster, I had opened up an interesting triangular space where the roof overhang carried on across the base of the dormer. I have always enjoyed searching out and putting to use dead spaces in the frame of a house. My wife Annie suggested making a plant shelf out of this one, with a skylight over it (photo facing page). Instead of being cut into the roof, the skylight glazing here replaces a whole section of it, from dormer wall to the eave. The only change this requires to Thallon's detail is that the head flashing becomes a typical sidewall flashing running up behind the shingles on the dormer wall.

Incidentally, though, with respect to the flashings, I have discovered since completing this work that I can eliminate what Thallon calls the "bottom trim" by ordering a metal edge on the corresponding edge of the sealed glazing unit. This exposes the clips, but has the advantage of reducing the capacity of the bottom trim to collect dirt and debris washed down off the roof. If you make this alteration, remember to adjust the inside dimension of the clip to allow for the added thickness.

I ordered the glazing as a single sealed unit that would span the three rafter spaces. Like the other skylight glazing downstairs, this unit was made with 6mm laminated glass for the inner layer and 5mm standard window glass for the outer layer.

*Gaining access—*The process of laying out and cutting through the roof is the same as for any opening. I usually drill a hole or drive a nail from inside at a specified distance from each corner, and as always, I think twice about how to finish the inside before cutting through to the outside. In this case, I wanted the drywall to wrap around the corner at the wall and butt into the rafter at the roof plane. The lookout rafters were in line with the wall studs, so this meant doubling up the lookouts at either end of the opening and adding blocking to the undersides of them (drawing facing page). Once having decided this, however, I waited until after I had cut the hole before actually doing the work. That way, I could punch my layout holes exactly where the cut was going to be (down the middle of the added rafter).

As with skylights elsewhere in the house, some of the finishing work was a great deal easier to do before the glass went in. As for the plant shelf itself, I reasoned that the first few inches immediately under the glass would not be particularly useful. Boxing in this area between the rafters also would make it much easier to finish the shelf surface later. I used some resawn cedar for the box to get something like the same texture as the old rafters, and painted them white.

The solid-looking lintel is actually a carefully faked-up beam (drawing below). I wanted the look of a beam let into the wall at either end, but I didn't want to go to the trouble of cutting the extra stud and doing all that reframing. Besides, I had some perfect material for the 3-in. beam— Douglas fir, skip-planed so that it still had a slightly rough texture in spots. Perfect, except that it happened to be a 1x12. So I swallowed my purist notions of Truth In Materials and slapped in the conventional doubled-up 2x4s, ripped and carefully mitered the 1x12, and cut the drywall around it. When I filled the drywall, I flat-taped and plastered right up to the wood.

I installed the red cedar sill, machined with table saw and router, after the drywall was painted. And when I had oiled the sill and carefully masked all finished surfaces, I tiled the shelf with tiles left over from the kitchen floor, stubbornly saved for years with just such an unknown purpose in mind.

The shelf functions well. It gets good light all day, but it is at its best in the afternoon when the sun has left the skylight above it and falls directly on this south-facing pitch of the roof. Going upstairs, we get a view of branches and sky beyond our neighbor's roof, and going downstairs the eye is caught and refreshed by a bright and leafy alcove. —A. S.

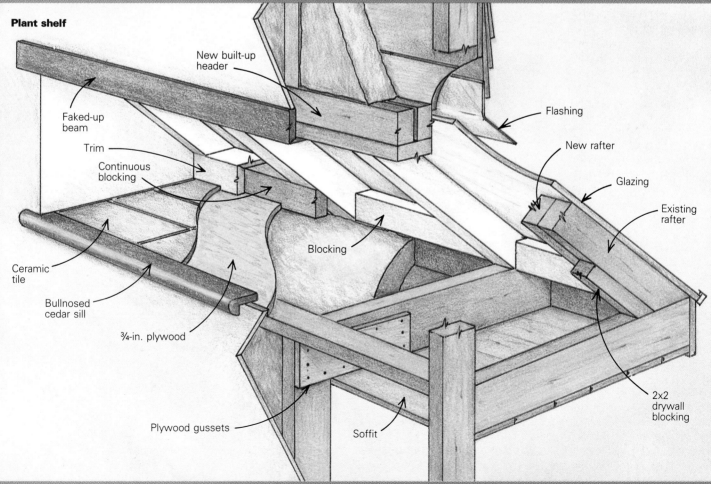

Plant shelf

New built-up header

Faked-up beam

Trim

Continuous blocking

Ceramic tile

Bullnosed cedar sill

¾-in. plywood

Plywood gussets

Soffit

Blocking

Flashing

New rafter

Glazing

Existing rafter

2x2 drywall blocking

either end wall. Not that a larger discrepancy from one end to the other would have mattered much; it would be visible only to someone looking up from directly beneath—someone lying on top of the piano, for example.

I am never able to quell entirely the prickling of apprehension at the moment I lower the blade of my circular saw into a perfectly good roof for the purpose of cutting a gaping hole in it. It seems such a perverse thing for a builder to do. But once I get going, goggles strapped on, earmuffs dulling the "chunk" of an old blade ripping through the roofing nails, the excitement always gets me. A hole in the roof dramatically alters the quality of light entering a room.

The length of the opening was determined by the location of the rafters. I cut right down the middle of them, or as near to the middle as would give me a relatively square opening. The uphill cut was in vertical line with the outside face of the original wall. The fact that this cut virtually intersected the bottom corner of a dormer wall gave me pause for a moment, but I removed some shingles and bent the old lead flashings out of the way, and then cut as close to the wall as I dared. The 1½-in. width of the outboard shoe on my saw seemed a good distance.

I had the skylight glass made up in two double-glazed units. This required a center flashing between them, which I considered a small price to pay for the ease of handling two smaller units. They were made with 6-mil laminated glass for the inside pane and 5-mil float glass for the outside pane. Manufacturers of sealed units recommend a maximum 1-mil difference in thickness between inside and outside panes.

Indirect light. **A trio of porcelain light fixtures, hidden from easy view, augments the natural light from above. The surfaces are painted with semi-gloss enamel to better reflect the light.**

Prefinishing the interior—I extended the line of the wall above the opening by installing 2x4s on both sides of each rafter. Then I placed, taped, finished and painted the drywall by reaching through the open roof. Working on the drywall from the scaffolding was distinctly easier than cramming myself into the constricted space between inside and outside walls of the bay with pieces of drywall, mudding trowel, pole sander and paint roller. I painted this important reflective surface, as well as the inside surface of the long wall of the bay, with a semi-gloss white latex paint. The degree of reflectivity could be increased even more by using aluminum foil to cover the portion of the wall above the header.

I also wired three simple ceiling-type light fixtures on the inside of the wall above the beam (photo left). I found that 60w "Fat Albert" bulbs provided a comfortable level of illumination, both as direct lighting for our hypothetical pianist and as indirect lighting for the rest of the room.

The inside of the long wall of the bay is paneled with 1x6 knotty shelving pine, milled down to ½ in. thick to tuck inside the corner of the window frame; the boards have a ship-lap edge to allow for contraction. All the imperfections in the boards were allowed to show, the approach being to treat it rather as though it were concrete formwork and not a slick, wood-finished interior wall.

The effect—With flashings and glass installed and everything painted white, the effect was much as I had imagined, though perhaps not quite as bright as I might have hoped. If anything, I would put the beam lower to emphasize still further the bay's alcovelike quality.

Instead of the threatened piano, we put a sofa there, and it is a pleasant place to sit, separate from the activities of the busy kitchen and yet just as easily connected.

During the summer months, the evening sun comes in the west window, and in late September you can sit there in the morning and momentarily catch the glow of the sun rising in the east window, as it clears the houses on the opposite side of the street. □

Tony Simmonds is head carpenter with Domus, a design/build firm in Vancouver, B. C. Photos by Charles Miller.

Building Louvered Shutters

Jigs and careful planning make quick, accurate work of a potentially tedious job

by Rob Hunt

A few years back, I was asked to build movable louvered shutters for a Victorian house that was built in 1888. The house's new owners were restoring the building, and they wanted new shutters made to match the old ones that had deteriorated over the years. Thanks to this first commission, we've been building shutters for a number of clients. Out of necessity we've found ways to make the work go quickly and accurately.

Many early homes had shutters. Hinged to open and close over windows, and with movable louvers, they served to protect the windows from bad weather and to diffuse the incoming light. Changing the position of the louvers changes the flow of light and air through the room, giving you a range of lighting conditions to choose from. Shutters can do a lot to enhance windows, and they give you more control in adjusting the amount of

daylight you want in a room, and more control over the ventilation day and night.

The frame of a shutter consists of two vertical members, called stiles, and at least two horizontal members, called rails. The shutters shown here have two central rails in addition to the top and bottom rails, dividing each shutter into three sections. The louvers are beveled slats with round tenons that fit into holes in the stiles. Each set of louvers pivots as a unit thanks to a vertical rod coupled to the louvers by an interlocking pair of staples.

Cutting the louvers to uniform size, calculating the amount of overlap (the spacing between each louver) and mounting the louvers accurately within the frame are critical steps in the production process, and we've been able to increase both the accuracy and speed of the job by using layout sticks and several jigs that I'll describe as we go along.

Layout—The first step in laying out louvered shutters is to make precise measurements of the opening where the shutters will go. Measure from the top of the window or door jamb to the sill on each side, then measure across the jamb at top and bottom and at several places in the middle. Taking extra measurements for height and width is especially important in older houses, since their jambs are seldom square.

If the jamb is out of square by ⅛ in. or less, I build the shutters to fit the smaller measurement. If the skew is worse than this, I build them ⅛ in. smaller than the largest measurement and then trim them to fit the opening after they are completed.

When you measure the height of the opening, remember that the sill is beveled. Run your tape to the point where the outside bottom edge of the shutter will hit the sill. Then

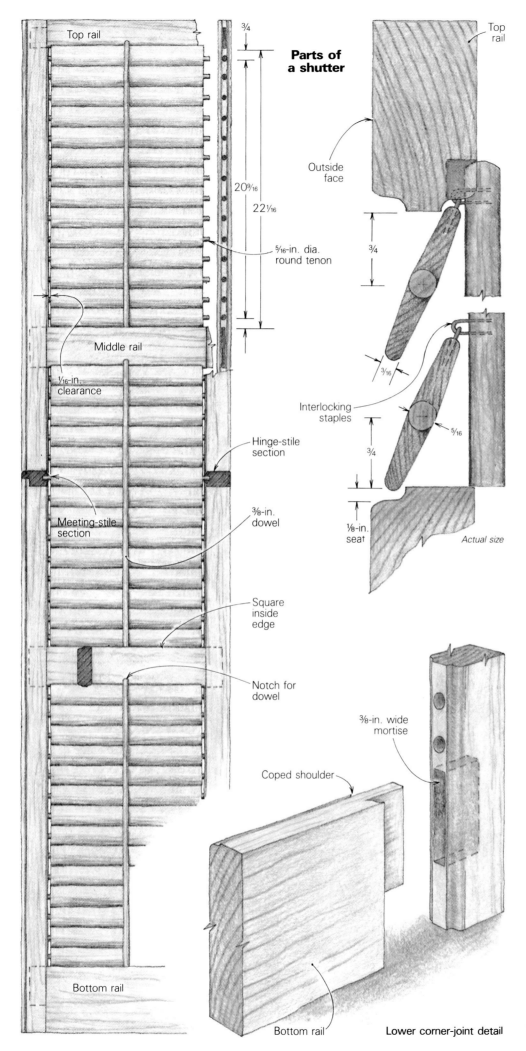

Parts of a shutter

Top rail

20⁹⁄₁₆

22¹⁄₁₆

5⁄₁₆-in. dia. round tenon

Middle rail

1⁄₁₆-in. clearance

Meeting-stile section

Hinge-stile section

3⁄₈-in. dowel

Square inside edge

Notch for dowel

Bottom rail

Top rail

Outside face

3⁄₄

Interlocking staples

3⁄₁₆

5⁄₁₆

3⁄₄

1⁄₈-in. seat

Actual size

3⁄₈-in. wide mortise

Coped shoulder

Bottom rail

Lower corner-joint detail

when the shutters are built, you can bevel the bottom edge of the rails.

For projects like this, which require a number of identical, precisely cut pieces, we find layout sticks very helpful. They serve as full-size templates and contain all the required dimensions and joinery details. The horizontal stick we made for these shutters is basically a full-size sectional drawing. It shows the width of the opening, the width of the stiles, the length of the rails, the louver dimensions, the bead on the stiles and the coped mortise-and-tenon joint that joins stiles and rails. All these measurements are critical, since each pair of shutters has to swing closed along a 1⁄4-in. rabbet. And finally, the louvers need to operate smoothly, without binding.

The vertical layout stick for these shutters is marked off to show the length of the stiles, the location of the rails, and the stile holes that accept the tenoned ends of the louvers.

The spacing of the louvers is determined by louver size and the degree of overlap desired, the overall height of the shutter and the width and number of rails per shutter. The shutters shown here are 79³⁄₄ in. high and have four rails. The top rail is 2 in. wide; the two center rails are 3 in. wide; the bottom rail is 5¹⁄₂ in. wide. Measuring between the top and bottom rails gives us 72¹⁄₄ in.; so to get three equal shutter bays and allow for the 3-in. wide central rails, each bay must be 22 ¹⁄₁₆+ in. high (the plus means a heavy ¹⁄₁₆ in.).

The next thing to figure out is the number of louvers you need, and the spacing between them (actually the spacing between bore centers for the holes in the stiles that receive the round tenons), so that the holes for the tenoned ends of the louvers can be marked on the stiles. Louvers on most Victorian shutters are 1³⁄₄ in. wide, and this is the width we used (Greek Revival style shutters have 2¹⁄₄-in. wide louvers).

The top and bottom louvers in each section must be located first, since their positions determine how much the louvers can close. Closing the louvers onto the bead is best for shedding water, but if you want to shut out the light, you need more pivoting clearance at top and bottom, which will allow the louvers to seat against each other. We centered the holes for the top and bottom louvers 3⁄4 in. from the top and bottom rails, sacrificing complete closure for a 1⁄8-in. seat against the bead on the rail (top detail drawing).

Once the holes for the top and bottom louvers have been located, we can figure the spacing for the rest of them. We want the louvers to overlap about 1⁄4 in. when closed to block the light and shed water, so the tenon holes in the stiles should be approximately 1¹⁄₂ in. apart, give or take a very small amount. The next thing that we do is to measure the distance between the top and bottom hole in each section (20⁹⁄₁₆+ in.) This we divide by 1¹⁄₂ in., yielding (thanks to my calculator) 13.713 spaces per section. You can't have fractions of a louver, so I divide 20⁹⁄₁₆ in. by 14, and get 1.469, or 1¹⁵⁄₃₂ in. between louvers. This is the spacing that I check by locating

Illustrations: Christopher Clapp

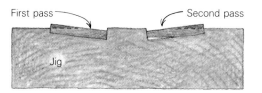

First pass — Second pass

Jig

Shaping louvers. Slats of white pine or cypress are beveled by running them through a jig (drawing above and photo at right) that's clamped to the planer bed. Grooves in the jig are angled at 6° so that the planer removes a narrow wedge of wood to make the bevel. Flipping the slat and running it through the other groove completes the beveling job. A second jig, far right, acts as both fence and stop as the squared tenons of each louver are rounded with a plug cutter. The shoulders of the tenons have already been cut on the table saw.

With a hollow-chisel mortiser, the stiles are mortised to receive rail tenons. Holes have already been drilled to receive the louver tenons.

the centers for the dowel holes on my vertical layout stick.

Another way to space your louvers is to adjust your rails so that 1½ in. will divide into the bay height equally. But for most of the shutters we've built, I've been matching existing shutters, so rail sizes and locations have been fixed.

Cutting louvers—With layout work done, it's time to make the parts. We start with the louvers. The completed louver needs to have two bevels on each side, and the long edges have to be rounded over. Each louver also has to be tenoned at each end to fit into the holes in the stiles.

To produce the slats for the louvers, we first take rough 8/4 stock—either white pine or cypress—and mill it down to a net thickness of 1¾ in. Then we rip strips ⅜ in. thick by 1¾ in. wide and surface 1/32 in. off of each side.

To bevel the louvers, we use a special jig, sometimes called a slave board, that we clamp to our planer bed. It's made from a piece of 1¾-in. thick oak that's slightly longer than the planer bed and about 7 in. wide. Two shallow, 1¾-in. wide grooves in the top face of the slave board are what make the jig work. The groove bottoms are angled 6° off the horizontal, and this slightly angled running surface for the wood strips lets the planer waste a narrow triangular section from the strip to create the bevel (photo top left). Each strip of wood is fed through one groove, then flipped and fed through the other, giving you all four bevels with just two passes through the planer, as shown in the drawing above.

To round the edges, we run the beveled strips through a 3/16-in. bead cutter on the shaper, using feather boards to hold them in place. Now all that remains is to cut the slats to length and cut round tenons on their ends. The shoulders of the louvers should clear the edges of the stiles by at least 1/32 in. or else they will bind when they are painted. We use a clearance of 1/16 in. to stay on the safe side.

We cut the louvers to a length that includes the ⅜-in. long tenon on both ends. Then we cut to the shoulder lines of the tenons on the table saw, holding the louver on edge against the miter gauge and setting the blade height to leave a ⅜-in. square tenon on both louver

ends. The tenons get rounded with a 5/16-in. plug cutter chucked in a horizontal drill press (Shopsmith). As shown in the photo top right, we use a fence to guide the louver into the plug cutter and a stop to keep from cutting into the shoulder.

Stiles and rails—We start with rough 6/4 stock, joint one edge and one face, and then surface it to a net thickness of 1¼ in. (frames for some shutters may be as thin as 1⅛ in.). The stiles for these shutters are 2 in. wide, and they're mortised to house tenons on the rails. We cut the mortises with a hollow-chisel mortiser (large photo above) and the tenons on a small, single-end tenoner.

The holes in the stiles for the louvers get drilled at this time too. Rather than mark and

drill the stiles directly, we've found it more foolproof to make a drilling jig. It's the same length and width as the stile, and the hole centers are transferred onto it from the vertical layout stick and then drilled out on a drill press. We attach the template to the stile with C-clamps, then start drilling, making sure the depth of the stile holes is ⅛ in. deeper than the length of the tenons.

The lower edge of each rail on the inside face of the shutter is left square, while the upper edge needs to have a short notch in its center to accept the vertical dowel. The notch and the square edge allow the shutter to open and close securely.

The remaining inner edges of the frame get an ovolo bead. This means that the shoulders on the rails have to be coped where they meet

the molded edges of the stiles. Coping is tough to do by hand, so we use a shaper fitted out with a three-wing coping cutter (a Rockwell 09-128 male sash) on a stub spindle. The stub spindle allows the rail's tenon to pass over it, so we can cope one shoulder at a time.

The last pieces to make are the dowels that will be attached vertically to each bank of louvers. We make these on a shaper with a ½-in. beading bit.

Assembly—Before putting everything together, we need to make the jigs that will hold the louvers in uniform position. The drilling template for the stiles can be converted to a louver jig by cutting ¼-in. grooves across the center of the holes. We make another jig just like this one and use the pair to hold the louvers. In the photo above left, both jigs are lying on the table while I test-fit stiles to rails.

Assembly isn't really that tricky when you use these jigs, though you have to work faster than the glue that's used on the mortise-and-tenon joints (we use West Systems Epoxy, made by Gougeon Co., 706 Martin St., Bay City, Mich. 48706). We first glue and assemble one side of the shutter completely, pressing all stile-to-rail joints home. Then we lay this sub-assembly on top of the jigs, apply glue to the exposed rail tenons, and engage them in their mortises. The trick here is to leave just enough clearance for the dowels at the ends of the louvers, as shown in the photo below left. Here's where fast work is important. Have your louvers ready, get them all engaged in their holes and set in the jig, then close the joints between stile and rails.

You can cut the dowel to length after stapling it to the louvers, or you can cut before stapling. The length of the dowel is the distance between the rails plus the length of the groove in the rail at the top of the section. Round the top of each dowel so that it will fit into the groove when the louvers close.

The dowel receives staples at intervals equal to the distance between louvers; and the uppermost staple on each dowel should be located far enough down the dowel so that the dowel fits into its groove in the upper rail when the louvers are closed. We use an Arrow T25 stapler with ⁹⁄₁₆-in. staples. It's a stapler that is used for putting up small wire, so the staples don't sink all the way in.

Next we attach each dowel to its section of louvers by shooting staples into the louvers through the staples already in the rod (photo below right). We learned, after some mishaps, that grinding the bevel off the staple points made them shoot straight in, with a minimum of splitting.

We usually leave painting the shutters to someone else, but it's important to seal the wood with a wood preservative before the finish coats are applied. Spray application is far better than brush-on because of the shutters' many movable parts.   □

Rob Hunt is a partner in Water St. Millworks and a cabinetmaking instructor at Austin (Tex.) Community College.

Assembling the shutters. **Above, test-fitting the mortise-and-tenon joints between stiles and rails before final assembly. The grooved jigs on the table will hold the louvers as the shutter is assembled, as shown at left. Rails are joined tightly to the right stile, but loosely to the left one, providing clearance so the tenoned louvers can be inserted in the stile holes. The final step, below, is stapling the vertical rod to each louver section. Double-stapling keeps louvers aligned with each other, so they can be opened and closed as a unit.**

Photos: Jane Hunt

Shutters from central Texas

Bright sunlight, long days and humid, hot weather made operable louvered shutters necessities for 19th and early 20th-century Texas houses in the San Antonio and Austin areas. The louvers let air in and kept rain out even on the foulest days, and on clear, warm days, they screened interior spaces from direct sunlight, while allowing for cross ventilation and natural convective cooling. Both air flow and ambient light levels can be regulated by adjusting the louvers. As the examples here show, shutters can enhance various architectural styles.

The photo directly below is a view from inside the window bay that's seen from the outside in the photo at bottom left.

Bay Windows

People think they know what a bay window is, but who can really define it? *Webster's* says it like this: "A window or series of windows usually rising from the ground and projecting from the wall, forming an alcove or recess within." But other languages take a different approach. Here in Holland the word is "erker," and the Dutch version of *Webster's* defines it this way: "Annex, wing, extension; an angular or round built-out section on a façade whereby a room springs out above the street." So you thought you knew what a bay window was? So did I, until I went looking for the perfect example.

—*Dennis Weisbrod, Bussum, Holland (photos by author).*

Sousse, Tunisia

Weesp, Holland

South Chailey, England

Bussum, Holland

INDEX

The articles in this book originally appeared in *Fine Homebuilding* magazine. The date of first publication, issue number and page numbers for each article are given at right.